TER
RIBL
E AD
VICE

Maxim Frolov

A PRACTICAL
GUIDE TO
EXECUTION
EXCELLENCE
IN TURBULENT
TIMES

EVERYTHING YOU'VE
BEEN TOLD ABOUT
SUCCEEDING DURING
A RECESSION IS
WRONG (AND WHAT
TO DO ABOUT IT)

Maxim Frolov
www.maximfrolov.consulting

Publisher's Cataloging-In-Publication Data

Names: Frolov, Maxim, author.
Title: Terrible advice : everything you've been told about succeeding
    during a recession is wrong (and what to do about it) : a practical
    guide to execution excellence in turbulent times / Maxim Frolov.
Description: [Moscow, Russia] : Maxim Frolov, [2020] | Includes
    bibliographical references.
Identifiers: ISBN 9781734957006 (paperback) | ISBN 9781734957013
    (hardback) | ISBN 9781734957020 (ebook)
Subjects: LCSH: Success in business. | Recessions.
Classification: LCC HF5386 .F765 2020 (print) | LCC HF5386 (ebook) | DDC
    658–dc23

Contents

Introduction:

Evolve or Go Extinct

Every career and every business look different. But all employment and self-employment scenarios have one thing in common. It's easy to get stuck. Frustrated. Lost. You keep doing the job required of you the way you've always done it—the way it's supposed to be done—but for some reason, you don't see the results you expect anymore.

That's when you search for advice outside your job or your circle. Motivational gurus and experts pelt you with inspiration on social media, at conferences, and all over the internet. It's easy to consume their exaggerated advice because it seems to be based on true stories. The path to success looks like a guarantee for you as well. Everyone needs some motivation at the right moment. But what's next? You're motivated to get rich, change the world, build your empire, and make history, but in reality, you don't know where to start or what to do after that. You're inspired, but what are the practical steps to take and the traps to watch out for? What works for those other than the top 1 percent, the most successful people in the world? What works to enable average people to succeed in business?

Most of the time, these questions have no answers. You're pumped up on motivation, only to see that energy vanish by the next morning. If you're making progress to build a billion-dollar business, you don't need any advice because you're on top already. If you're not, you have questions specific to your unique situation. It's critical that you build a bridge between generic inspiration and specific action. Because the world doesn't give you the supplies you need. The majority of success advice you receive is awful because it's made to resonate with a wide audience.

Everything seems easy on social media. Experts sell you ideas, but ideas are overrated. Few can turn those ideas into tangible results like profit or sales revenue growth. They're missing useful details, ultimately misleading people to make wrong decisions and experience the harmful results—from poor performance to lost sales to even a failed business venture.

It doesn't seem like success should be this hard. The last decade brought us consistent global economic growth. The years 2009 through 2019 witnessed the longest bull market in stock market history. In the United States, the S&P 500 rose more than 370 percent over that ten-year period.[1] Wages have risen across countries and industries.[2] The unemployment rate in leading economies is at historic lows today. In the US, unemployment rates have fallen to their lowest in fifty years.[3] The stock market reached an all-time high on January 17, 2020.[4] The number of start-ups valued at $100 million or more is incredibly high. These and other observations imply that business is easy. This ease is an illusion. It's only a short matter of time before the next financial reset happens. The average length of a growing economy is just over three years.[5] Ours has lasted more than ten. We're running

---

[1] Anna-Louise Jackson, "10 Ways the Last 10 Years Were Remarkable for the Stock Market," Acorns, January 6, 2020, www.grow.acorns.com/stock-market-remarkable-decade-in-10-numbers.

[2] Ben Casselman, "Why Wages Are Finally Rising, 10 Years after the Recession," *New York Times*, May 2, 2019, www.nytimes.com/2019/05/02/business/economy/wage-growth-economy.html.

[3] WhiteHouse.gov, "Unemployment Rate Falls to Lowest Level in Nearly 50 Years; US Economy Adds 263,000 New Jobs in April," May 3, 2019, www.whitehouse.gov/articles/unemployment-rate-falls-lowest-level-nearly-50-years-u-s-economy-adds-263000-new-jobs-april.

[4] Kimberly Amadeo, "Dow Highest Closing Records," The Balance, March 20, 2020, www.thebalance.com/dow-jones-closing-history-top-highs-and-lows-since-1929-3306174.

[5] Cameron Keng, "Recession Is Overdue by 4.5 Years, Here's How to Prepare," Forbes. October 23, 2018, www.forbes.com/sites/cameronkeng/2018/10/23/recession-is-overdue-by-4-5-years-heres-how-to-prepare/#6128214440d8.

on steam. Throughout the last hundred years, an economic reset has happened every five to seven years.[6] Since the Great Depression, the global economy has endured thirteen recessions.[7] The coming decade more likely than not will bring number fourteen with it. Expect many years of volatility that will raise the bar for every business and every professional to be ruthlessly competitive and insanely efficient.

We've already seen a harbinger of things to come. In late 2019, a new virus unleashed itself upon the world. The coronavirus pandemic, causing a deadly respiratory disease called COVID-19, brought entire economies to a screeching halt. Add to that travel bans, citywide quarantines, business closings, mass migration to remote work, and major events canceled everywhere. Some have called it the Great Panic of 2020, others the apocalypse.[8] On March 12, the Dow Jones Industrial Average had the worst single-day loss since Black Monday back in 1987.[9] By April 2020, the United States saw thirty million job losses.[10] Estimates suggested over 100 million losses in travel and tourism alone

[6] Scott Galloway, "Every 7 Years," Seeking Alpha, July 31, 2017, www.seekingalpha.com/article/4092670-every-7-years.

[7] Dan Barufald, "A Review of Past Recessions," Investopedia, February 3, 2020. www.investopedia.com/articles/economics/08/past-recessions.asp.https://www.investopedia.com/articles/economics/08/past-recessions.asp.

[8] Sudeep Reddy and Victoria Guida, "The Great Coronavirus Panic of 2020," Politico, March 8, 2020, www.politico.com/newsletters/morning-money/2020/03/09/the-great-coronavirus-panic-of-2020-785941.

[9] Jessica Menton, "Dow Caps Week of Wild Swings by Soaring over 1,900 Points as Trump Frees Up Financial Aid to Fight Coronavirus," USA Today, March 13, 2020, www.usatoday.com/story/money/2020/03/13/dow-stocks-poised-open-higher-after-worst-day-since-1987/5040329002.

[10] Katia Dmitrieva. "Job Losses Deepen in Pandemic With U.S. Tally Topping 30 Million," Bloomberg, April 30, 2020. www.bloomberg.com/news/articles/2020-04-30/another-3-8-million-in-u-s-filed-for-jobless-benefits-last-week.

due to the coronavirus.[11] In May, the International Labour Organization projected that half the entire global workforce was at risk of losing their income—that's 1.6 billion people.[12] The Great Depression was a speed bump compared to the new normal forced upon us all.

Chances are, at the moment you're reading this book, the economy is nowhere near its January 2020 health, nor are we. Perhaps you are reading this book in self-isolation either to wait out the two weeks it can take to exhibit symptoms of the virus, or you are protecting yourself from others who are. Whatever your present circumstances, you are most likely worried about your safety, your job prospects, and your personal financial situation as we wait to see just how bad the situation will get.

If you want to survive a world in chaos, you'd better get yourself ready. Most advice out there will not help you do that at all. As the economy hits the wall, every business is starting cost-reduction exercises. The least efficient employees always go first. The least robust and profitable businesses die first. It's the law of natural selection. In a situation where there is no bounce-back trend in mind (like when the S&P 500 gained back 84 percent during year two of the Great Recession), execution excellence is crucial to business success.[13] What do I mean by *bounce back*? In most recessions, especially during the most recent one, we saw a bounce-back effect. There was a debt default scenario, and the

---

[11] Olivia Sharpe. "WTTC estimates over 100m job losses in travel," Cruise Trade News, April 27, 2020. www.cruisetradenews.com/wttc-estimates-over-100m-job-losses-in-travel.

[12] International Labour Organization. "ILO: As job losses escalate, nearly half of global workforce at risk of losing livelihoods," April 29, 2020. https://www.ilo.org/global/about-the-ilo/newsroom/news/WCMS_743036/lang–en/index.htm.

[13] Katharina Buchholz, "How Fast Has the Economy Recovered After Past Recessions?," Statista, March 17, 2020, www.statista.com/chart/21144/s-p-500-recession-recovery.

stock market crashed, but within the next couple of years, governments and markets applied the necessary fixes. We bounced back. Then unstoppable growth occurred. The bear became a bull. The stock market saw its highest single-day increase since 1974.[14] But based on the view of many smart, trustworthy economists and finance market professionals, the coming recession will be very different. It might not even be classified as a classic recession. Most likely we'll see a rushing-hills scenario with lots of ups and downs, quarter by quarter; one quarter up, one quarter down. Don't expect your business to bounce right back. Expect a roller coaster.

In this new economy, getting paid well and earning a great living is much more difficult. Economic circumstances are applying massive pressurc on the cost of doing business around the globe. On top of the panic and the pandemic, disruptive technologies like artificial intelligence will keep cutting jobs for the sake of efficiency. This will create massive opportunities for those who are prepared and risks for those who are not. Then it will be time to evolve . . . or go extinct.

Winning Long Term Requires Practicality

This book accumulates practical advice that makes businesses and careers successful in real-world situations. These situations are common and applicable to the majority of businesses and geographies. This advice is based on true business cases I've had the pleasure to be involved in. Throughout my career, I've had a chance to live, travel, and do business in more than one hundred countries. Every country has its own specific customs. How many times have you heard the following statements: "Yeah, but our

---

[14] Claire Ballentine and Vildana Hajric, "S&P Caps Best Week Since 1974 after Fed Acts: Markets Wrap," Bloomberg, April 9, 2020, www.bloomberg.com/news/articles/2020-04-08/asian-stock-futures-rise-after-s-p-500-rally-markets-wrap.

market is different. This is what you've seen and used, but it's different here." That's partially true. Every country and every economy has its own ways. Yet while business dynamics differ across geographies, business productivity and efficiency metrics are the same all around the world.

Throughout my career, I have held very different roles and responsibilities, from being a one-man show in a start-up taking a shot to disrupt the world to managing sales worldwide for global corporations trying to survive massive scandals and geopolitical pressure. For most of my career, I've done nothing but firefighting, fixing, crisis managing, starting up, and starting over. I've had to be practical about the tasks I've dealt with, had to have an entrepreneurial mind-set, and had to have taken full responsibility for results. My career path has required relevant skills that evolve every single day.

I was lucky enough to pick the technology industry at the beginning of my journey, as it helped me to become a global citizen. Global citizenship was all I wanted from the beginning of my career: to live where I want without borders and to seize exciting business opportunities. Holding executive-level roles in companies of different sizes, from the largest corporations of the world to the smallest start-ups, I've gotten a chance to learn from my own failures, build scalable businesses, and help thousands of people become better professionals and better people. While doing this, I have become a better person as well.

I wrote this book for businesspeople who do not necessarily want to build the next SpaceX, Apple, or Facebook but who *do* want to be successful, make money, pay bills, have better living standards, and be proud of their work. I also aim to help those who've gotten stuck in their careers remove the boundaries with practical advice and take the next step forward. The case studies and pragmatic strategies included in this book, which are based on debunking popular (read: wrong) advice, are here to help you. In the chapters ahead, we will cover the following:

- How to turn any underperforming business or project around

- How to leverage the worst-case scenario into recession-proof strategies

- How to decide which metrics to measure so that success is all a matter of math

- How to achieve any goal without moving the goalposts

- How to change course the smart way so that you don't lose progress

- How to earn the income you desire

- How to play on the biggest business stages without faking it

- How to hire slow and fire fast without ruining relationships

- How to performance-manage a company out of a crisis

Whether you're an executive, a manager, or an employee at a Fortune 500 corporation, or whether you run your own business of any size, this book will offer the actionable, step-by-step answers you seek so that you can stay relevant to the market, outsmart the competition, and win in the next decade and beyond. Let's begin.

CHAPTER

Stop Talking and Start Doing:
Terrible Advice about Quick Wins
(And What to Do Instead)

Have you ever heard advice like "Stop talking and start doing"? Sounds useful, doesn't it? Better to take action without a perfect plan than to delay a minute longer, right? That's why everybody is always looking for the newest success tip about how to quit procrastinating. Imperfect action is preferable to perfect preparation. This is what many business leaders, managers, and entrepreneurs are committed to doing—acting for the sake of doing something in the hope that the effort will create positive momentum.

It's terrible advice. Think about what it is that you want to do. Increase your sales. Hire better people. Transform your business. Find a rewarding career. Whatever it may be. If you are going to commit to the result you want, why would you not take the time you need to precalculate each step? Why would you throw yourself into execution without knowing *how* you're going to execute when things go wrong? Why would you stop talking and start doing?

Proper planning is not procrastination. It's preparation for success. Success exactly as you imagined. Have you ever heard the saying,

"An idiot with a plan can beat a genius without a plan"? I know a few things about this. Let me tell you a story.

The Casual Vacancy: Not Your Typical Business Transformation

December 19, 2016. Less than a week before Christmas and fewer than two weeks before the New Year. Four thirty in the morning in Los Angeles, California. A conference call with Dubai, Istanbul, and Johannesburg has been underway for the last hour and a half. Fiscal year closing forecast and Q1 2017 readiness are done. I ask my general managers (GMs) how they feel about passing the point of no return. Did they believe four months ago that they were taking a risk by leading the largest business transformation in company history?

Their answers are too much information to process this early, but we are still OK with the timeline. We don't really have a choice. Managing a business from the other side of the planet in real time demands quite a routine. It's gone on like this for the past two months.

It's worth it.

In summer 2016, Kaspersky Lab asked me to turn around their lowest-performing region globally—the Middle East, Turkey, and Africa, or META for short. About sixty countries in total. At the time, Kaspersky Lab was all over the news because of a huge scandal in the United States. Those headlines would break in just six months, but at the time, company leadership had other concerns.

"We need someone with your experience and expertise to fix things in META," Kaspersky's chief sales officer (CSO) told me during the first minute of our meeting at Kaspersky's headquarters in Moscow. "What we need is a new managing director over the region to make some changes."

It was like a blast from the past. We had had a similar conversation in the same office in January 2014. I was offered a position managing emerging market (META, Latin America, Central and Eastern Europe) territories with similar goals. Fix underperformance and clean house.

"I appreciate the offer, I do. If it had come at a different time, perhaps, but . . ." I trailed off. "I'm in the process of exiting my start-up, Urban Innovation Group. And also, I'm moving to be with family in Los Angeles at the end of October. Our baby is due right before Christmas. No travel outside the United States until January. I don't know if starting a new position before moving to the other side of the world is a great idea."

"You can work remotely if necessary and be back in Dubai in February once you are able to."

"What exactly are these problems?"

"Management is out of whack. Bottlenecks are everywhere. Low sales."

"Even if I started next week, I would have only a few months to build a plan for you, sell that plan to employees, and transform the team, the sales process, and the business model."

"It might not be as complicated as all that. You've done it for us before. Just put together a team and see what you can do. Get the region back on the horse, so to speak."

I accepted the position. It would be my job to take over a failing business and turn it into a solid profit generator. What an opportunity. Kaspersky Lab is a global player in the cybersecurity industry. Headquartered in Moscow, Kaspersky has sold famous cyber protection products all around the world for over twenty years. Back in 2016, cybersecurity was already an overcrowded space, with more than a thousand competitors globally. Competition rose as new players kept entering the market. Kaspersky's bread-and-butter product, endpoint security, was fast becoming a commodity with no unique value proposition (UVP).

An outdated business model made my mission look even more impossible. Kaspersky operates five lines of business, from consumer to enterprise, doing both direct to consumer and distribution through resellers. The company's past success was driven mostly by the consumer and small business sales, segments where the cost of doing business was constantly increasing. But the average deal size (ADS) had almost no chance of increasing unless we had a unique proposal that differentiated us from the large number of competitors. Unfortunately, that was not an option for Kaspersky. Previously great products had been commoditized quickly. This business model had its complications for the managing director role, as I had to orchestrate several businesses simultaneously. Add the geography of more than sixty countries within the Middle East and Africa, with all their diversity, cultural and business specifics, and economic and political circumstances, and my decision to accept the job offer wasn't easy.

I was up to the challenge. I saw an opportunity to accomplish something that no one in this company had ever done—change the business model on the go and turn the underdog into the hero. When I officially stepped into the position on September 1, I was aware that the situation wasn't good. But no one was able to tell me what was wrong, why they struggled to deliver sales numbers, or what they'd tried before to fix the situation. Besides the CSO's vague advice about "some changes," no company leader could tell me what they expected from me. They all told me what they *didn't* want—big changes or disruptions. To me, that was a clear indication that they were stuck in the past with their understanding of what the market situation was and what makes a business model successful long term. When your sales are falling to the level where you face negative earnings before interest, taxes, depreciation, and amortization, the last thing you expect is a business model change. You have broken with reality.

"This region might be slipping right now, but we've had a glorious past. So much value. Big operational model changes could hurt us," said a board member during my first week as managing director. "What we need are fast fixes. Quick wins."

"I understand," I said.

Remember how Charlie Harper of *Two and a Half Men* said, "I understand" so often? In a later season, he explained that "I understand" means "Yes, I'm hearing what you're saying. It doesn't mean I understand or agree."

I was Charlie Harper that day.

Quick fixes make everything worse. They accelerate a business's free fall in most cases. They fix the symptoms of the problem, not the root cause. Fast fixes help people accomplish the CYA (cover your ass) exercise and report the problem as solved, thereby easing management pressure. So I did the opposite of a quick fix. I ran a full-scale business model transformation, including consulting-level assessment and a long-term strategic growth plan that involved partners, employees, marketing, sales—everyone. A complete overhaul. This meant fourth-quarter 2016 sales targets would go unmet. However, the bottom line still had a chance for a positive outlook. The region wouldn't become a money-making machine until January 2017.

This is what I told company management while selling the business model evolution proposal. "Money-making machine" made everyone smile. Of course, they'd heard that promise so many times before from previous managing directors. The difference was, I was serious about making the transformation work, and I would be there every step of the way to ensure that it did.

After selling the proposal, I held an all-hands meeting with the META leadership team. I briefed them on the plan to transform the region, not to go for quick wins. I then locked myself in my office with four other stakeholders, including a business-to-business (B2B) director, the regional finance director, and two business intelligence (BI) analysts. We got right to work looking for the root cause of the massive underperformance.

We finished our in-depth business assessment and transformation plan in three weeks. I presented the plan to company management and shared the root-cause assessment—why the current business model was broken in the first place. I explained what needed to be done, how long it would take, the risks, and the opportunities.

"The transformation will be painful," I said. "Don't expect to be wowed with immediate results. A lot of people will be pissed off. It will take a solid three months to execute the plan and even longer to see results. Based on this plan, I expect to double revenue in four years." I ended my presentation with our clear objective—start to deliver results in January 2017 and be consistent every single month onward. The region would be more successful than it had ever been. It was an ambitious plan that was accepted with skepticism, but it was one that I knew I would accomplish.

What made me so confident in my ability to pull this off? Self-awareness is my foundation. I know what I am capable of and what I am not. I know how to use my strengths and to deal with my shortcomings. I knew I could execute the plan because I'd already accomplished similar feats in previous businesses. I was patient enough to develop a plan A, a plan B, a plan C, and even a plan Z. I had enough empathy to win employees' hearts one by one, as I was aware of the turbulence I was about to create. I would get people to follow me because they would understand why we're doing it, how we're doing it, and what the shiny prize at the end would be.

The plan was approved. The game began.

On September 23, 2016, I held a one-day workshop to announce the business plan to the whole staff. Everyone attended in Dubai, Istanbul, Johannesburg, and Moscow or remotely via teleconference. My conclusion that an outdated business model and inefficient business processes were the root cause of our problems surprised everybody. Everyone from regional leaders to new interns assumed that they weren't performing well because of some external factor. Of course, there *was* an external factor. It's called competition. Our competitors were beating us because they had better marketing, superior customer service, more skilled and qualified sales teams, or, most importantly, consistent business processes that connected the dots. On top of all this, our product portfolio had gaps in our value proposition (i.e., why you should buy from us?). Kaspersky was losing market share across the region. Competitors had everything META was missing, but we'd built a plan to find it.

We prepared a detailed road map with dates and deliverables so that everyone would know what to do and when. That would come in handy when I moved to Los Angeles in six weeks—a twelve-, eleven-, and ten-hour time zone difference from Dubai, Istanbul, and Johannesburg, respectively.

Not all three regional GMs bought into my recommendations at first sight. The Middle East GM was a new hire, so he aligned with the new direction without complaint. The GM over Africa had been in the role for a while and understood that the business model didn't work anymore, so he was happy to join the journey. The GM for Turkey wasn't as easy to convince. He blindly believed that what he was doing was right despite the fact that his annual profit and loss (P&L) results had waved a red flag for a few years in a row.

Not easy, but not impossible. I ultimately got him, along with his peers, to agree on the road map, the roles, and the expectations. We visualized who did what on every level, going down to each employee and their competence. That was vital. It was the only way we could make any process work without babysitting (i.e., micromanaging).

The GMs' teams had a challenge digesting the transformation plan we fed them. Of course, new plans cause fear. No one likes change, even when it's good. That's a human being thing. New plans meet skepticism and silent resistance to some extent. What helped to remove the roadblocks was to patiently guide employees through our big plan, to be granular with every explanation, to let them visualize themselves following it, and to understand and feel that it's sensible and doable. In the end, deep understanding of what this game was about led to the most important thing our team needed—*confidence*. Playing with confidence maximizes the chances of winning. Ask any sportsman. Building confidence among a group who's been losing for years is crucial to execute any plan.

With confidence came commitment. From September 16 through October 1, we completed a sales and marketing resources assessment. We built a framework to understand whether every team member had the skills required for the new business model, whether they could be developed fast, and if not, whether there was potential. We made it fair and precise. Every conclusion was based on numbers and business-segment-specific metrics. For example, while assessing enterprise sales managers, we did not look at their deliverables during the last few years. We focused on new revenue generated, the number of complex deals they did per year, average deal velocity and win rate, customer retention trend, strategic product mix, and profitability. How often did they discount, and how deep did they go with them? Could they deliver a winning value proposition without building that UVP on price?

The channel roles assessment—the people who manage distributor and reseller performance—was the toughest part of the exercise. The company has a two-tier distribution model. In two-tier distribution, resellers purchase products through their distribution partner, not through the company. Managers looking after the partner channel have a lot to orchestrate and influence. Numbers-based decision-making helped us decide who stayed and who left. Metrics told us more than any story we had been given.

By October 1, it became clear that required organizational changes were bigger than I expected, especially in the Middle East, the region's key sales contributor. We had to replace about 60 percent of the salesforce. Most current managers were not able to play the enterprise league game we wanted to enter. That meant we had to find better people. Back in 2016, Kaspersky Lab didn't have the best human resources brand in the Middle East. To convince any enterprise sales champion to join the company seemed to be a serious obstacle.

"I see our recruitment team struggling," I told the region's GM. "Candidates are not excited about working for Kaspersky."

"Let me lead the team in the field," Middle East General Manager Amir Kanaan said. "Lead by example at every stage of the enterprise process—account planning, joint selling, pipeline review, and sales skills development. I will act as an enterprise director of the region as well. This will help set the new higher standards that ultimately make us an attractive company to join."

With the GM retraining the enterprise sales team, I interviewed like crazy—three or four new candidates every day. I looked for people who would bring needed expertise to the company and help us build an all-new sales culture. A culture of winning in the big boys' league. I wanted only the best people who believed in my goal to turn the region around.

To me, "best" means what you *can* do, not what you've done. A lot of companies have a fundamental hiring problem. They hire and promote based on employees' past achievements (e.g., a shiny résumé), not on what they bring to future missions. Bad hiring kills a lot of great ventures and initiatives. We could not afford to hire the wrong people. The business culture we wanted to build was based on process simplicity and execution excellence at every step of the way.

We had to build processes on the go. From scratch. Having no proper customer relationship management and BI tools ready yet, we decided to build the productivity tools in Excel to fulfill our needs. They might look ugly, and they're not as sexy as modern BI tools, but who cares? The spreadsheets worked. We knew BI tools would come later, so we were happy to have a role model for numbers-based decision-making. Understanding the BI team's role in helping us achieve the big goal to double revenue assisted me in winning the hearts of the BI team at company headquarters and prioritizing META's tasks.

Numbers-based decision-making became our philosophy. Execution excellence became our religion. To make this work for teams who had never used intensive data to boost their productivity, we decided to take every employee in the Middle East, Africa, and Turkey back to school. We have developed training programs to help salespeople manage their sales and their partner and customer relationships more efficiently and to follow data and processes, not instincts like they always have.

We also had to focus on fixing our sales infrastructure and partner ecosystem. Most vendors in the technology business choose the two-tier distribution model because of scale and market reach capabilities. But it works only if a partner program, clear process, discount transparency, marketing support, and commitment are there.

We approached our distributors with a request for proposal (RFP). Our key objective was to assess our distribution channel using the same framework, reach alignment in goals and priorities, and optimize the distribution pool accordingly. We briefed our distributors on our goal to double the business in four years, shared our strategy, and demonstrated our commitment to executing this plan. Why an RFP? We wanted them to submit their vision of the partnership translated into simple business language. To make it fair and easier to submit, we created a simple yet comprehensive template for the business plan we wanted them to provide. The beauty of this tough exercise was our full transparency. We offered equal opportunities and support to all existing distributors as well as to a few candidates who were interested in starting a partnership. And it clicked. We were able to make strategic decisions based on the relevant information only, removing any bias. Of course, the process was an endless firefight due to the RFP, but it was all right because every stakeholder on my team was ready for it—we had trained on what to do in every situation that we had modeled in advance.

Once all proposals were submitted and reviewed by a diverse group of stakeholders (we called it a committee), we made the tough decision to replace more than half of our distributors. They were not able to help us with the market reach goals or the strategic enterprise sales development we wanted. We'd had a glorious past together, but evolution is a cruel thing. You've got to be relevant to the market today and have potential for tomorrow, not yesterday. Decisions were made, and terminations and onboardings were approved.

On October 28, I moved to Los Angeles to manage META remotely. Our plan's execution was in progress. I was happy with the way we stuck to the road map and schedule. However, moving to LA in the middle of the business model transformation worried me. I'd already prepared my GMs and other stakeholders for our changing relationship. We had developed and tested simple routines before the move. For one, I made my online calendar visible to everybody. Every single minute of my time was managed. Despite the fact that I was expecting a new family member, work-life balance didn't mean I was going to sacrifice our big task. I knew that I would eventually find a balance that fit both my family and my team.

That balance was a multidimensional equation. Every day, I woke up at 3:00 a.m. Pacific time, which was 3:00 p.m. in Dubai, 2:00 p.m. in Istanbul and Moscow, and 1:00 p.m. in Johannesburg. My workday started at 3:15 a.m. My first conference call was with my personal assistant for fifteen minutes. She briefed me on meetings for the day, who wanted to discuss what, what was ready from my previous day's task list, and what was not ready. Next were one-on-one conference calls with my GMs. They were to report their progress on tasks, update us on the transformation road map, and bring their problems to the table to discuss how we would solve them. These daily calls kept us all aligned—essential to tracking with the business model transformation plan. Every minute of my day (or night, to be precise) was used with a purpose. Zero waste was allowed.

My full workday lasted until 11:00 a.m. Pacific, when my team switched off. After that, I summarized everything we'd completed and prepared notes for the next day. Then, believe it or not, family time. My wife and I had a late breakfast and spent the whole day together. We lived in Calabasas, so Malibu was about twenty minutes away. I took more beach walks in those three months than in my whole life before. It sounds surreal when I write about managing extremely complex business processes in the opposite time zone. Meanwhile, I had to keep my wife constantly happy, making her feel loved and preparing for the baby's birth. Every day, she was as equal a priority as my 3:00 a.m. business.

Every day around 6:00 p.m., I locked myself in the home office, looked through my morning's notes, and added anything that had popped into my head during the day. I shot my notes to my GMs and my assistant to make sure that they had them when they started their workdays in a few hours. Then I went to bed around 9:00 p.m. Well, dropped and switched off, actually. When the alarm sounded at 3:00 a.m. the next day, it started all over again.

I kept up this routine for three months. Did I like it? No. I'm the one who complains about waking up at 6:00 a.m.! Yet I'd get up exceptionally early, drink a lot of coffee, grab my headphones, and start my day. I had a plan that I needed to execute, my commitment to it, and my military-forged discipline to keep this commitment.

Real commitment is a personal choice. It's commitment to the word you gave in front of people who believe in your leadership. It's commitment to work ethics and professionalism when nobody is watching. While forcing myself out of bed at 3:00 a.m. every day for three months, I knew I was not the only person sacrificing their comfort to make a difference. My team did the same on the other side of the planet. For example, my GM in Africa, Riaan Badenhorst, was leaving home at 6:00 a.m. to take his oldest kid to school. He rushed home every evening to help his wife with their newborn twins. Once the family job was done for the day, he locked himself in his home office to discuss with me the tasks his team had worked on. He and all other stakeholders worked until 11:00 p.m. every night to get the maximum share of my time. The same happened in Moscow, where the new B2B META director, Kirill Astrakhan, was working sixteen hours a day with no weekends to make sure the ball of new sales started rolling. No one complained about work life-balance ever. That's a real commitment.

The closer we came to the new year, the more time I invested in talking to my extended leadership team. There was a lot to articulate, as many of them were not mentally prepared to become doers who deliver starting January 1.

I wanted this team to know that they were winners, not based on my belief in them alone but based on their accomplishments over the last four months. I wanted them to see how big a transformation we had accomplished, how tasks that had looked impossible four months back were now done. They were set up for exponential growth in the new year. All because we'd built a solid foundation.

By December's end, we'd accomplished the mission. We'd made well-thought-out decisions about who was staying and who wasn't, what new partners and distributors we needed to recruit to make a difference in our revenue and secure enterprise business development, what processes every team would follow, and which metrics we'd use to measure and manage progress. My last message to the team in 2016 was "No panic. Our plan works." This became our mantra for 2017. Every all-hands call closed with "No panic. Our plan works." And it did work.

On January 31, 2017, it was a different business. We were well positioned for about 30 percent year-over-year (YoY) growth (which we did achieve that year), grew strategic enterprise sales faster than any other territory, built a strong distributor and reseller framework, and boosted team morale, resulting in almost zero voluntary turnover. We were a different group of people. Something more than an employment contract and sizable earnings expectations united this team—a big idea.

When I flew back to Dubai in early February, I saw a team working as one. Business processes functioned almost like a machine. The transformation went perfectly. It was a manifestation of everything I had wanted. It was official—we had planned and executed the largest transformation in Kaspersky's history on the go. We'd changed the business model, the partner ecosystem, the business processes, and the organizational structure across more than sixty countries. Most of our processes evolved into global standards within the next couple of years. The Middle East and

Africa remain the most predictably successful and profitable region for the company worldwide.

I have to admit it: I love sharing this story. Because I was set up to fail. When I took the position, no lines of business were delivering on sales targets. All of them declined year over year. Customers churned. Teamwork was awful. The partner ecosystem was out of control. The margin was such a question mark that the top leaders in headquarters wondered if they would even carry on doing business in that region. Before the transformation, META had been the least profitable, most problematic region. It would have been easy to give excuses and apply quick fixes. But we made a choice to give no excuses. We made a choice to execute with excellence instead.

Bottom line: we transformed a business that no one believed could be fixed—and so can anyone. Whether you're working with a team of five or a corporation of five thousand, you can make positive changes by implementing the following seven simple tips.

Tip 1: Identify the Problem

Always be as specific as possible in everything you do, especially when you're identifying problems. In most cases, problems are masked. This leads to a situation in which you are trying to fix not the problem itself but its reflection.

Selling software with ecommerce is a good example. If you run your ecommerce business and see that sales numbers are below the target and that YoY growth is negative, what would you do? Probably invest in web traffic to bring more prospects and offer aggressive discounts to get them to pay attention. This might work. But this usually has a limited impact on sales today and harms long-term results. The best-case scenario would be when you hit your targets and walk away from the problem till the next fiscal quarter. Then what? You'd likely face the same situation because the sales performance problem is still there. Something is broken, but you expect the engine to perform anyway.

If you quickly analyzed your revenue, you'd see a snapshot of which part of your business was in trouble, and you would be able to act more efficiently. This is not that complicated. Here are the steps:

1. Break down sales to new revenue and recurring revenue to see which is declining. Understand your renewal rate trend and the YoY sales trends of each.

2. Check for customer churn. What is the YoY trend showing you?

3. Check whether your average deal size is declining. Do it for each segment, new and recurring. Understand YoY trends.

Once you've accomplished these three steps, you are close to the reason why sales are below the target. As soon as you have identified which segment of revenue is in trouble, you can verify the root cause and apply the fix that makes sense.

For example, let's say that you've found out that your renewal rate is declining—that is, customers are renewing less than expected and your ADS has decreased. Ask yourself if an aggressive discount is the right response to your problem here. It would be very unlikely. There is a chance that overdiscounting is causing your ADS decline, and all the while you have customer experience problems, which would explain why a lower price couldn't buy their loyalty.

Whatever your situation, you have to get to the root of the problem. In this case, that problem likely lies in the customer experience. A simple BI outlook can give you enough insights to decide what to do next. You just need to ask the right questions.

Success in any industry today is impossible without data-based decision-making. If you don't have relevant BI tools, you have to build them. Such tools are rather simple to build and use. Pick the one that fits your business specifics. More importantly, learn how to read the data—that is, how to convert key performance indicators (KPIs) and metrics into the specific details of your business. Educate your team to do the same.

Never, *ever* skip these numbers. In a rush to get to work, you only set yourself back. I took my sweet time making unbiased decisions without influence from charisma or circumstances. I wanted to build a plan based on a 360-degree look into the situation, a plan based on metrics. That's why I locked my team in a room for three weeks. We saw nothing but the numbers, which always tell the truth. No emotion, no bias, no bullshit.

You can make the right decision based on relevant data only if you understand how to read it. Most people have a rough idea of how to interpret data. I've seen BI teams as big as twenty people generate beautiful reports that give you X-ray-like looks at where your business is today. Yet the conclusions their senior leaders came to would shock you. The report was perfect, but leadership did not know how to translate data language into business language. The key is not just to read the data but to know how to convert the data into action. Take your time to track the metrics and then take more time to understand it.

So many leaders have the world's most powerful BI tools at their disposal, but they struggle to read and interpret the data. To me, it looks like a *Wheel of Fortune* player guessed all the letters but somehow could not read the word.

If you're asking a common question like "Why are sales below the target, and what should we do about it?," always look at revenue by the numbers. If you're in enterprise sales, break revenue down by account manager, customer segment, and geography. Know your historical win rate. For most businesses, 30 percent is a pretty solid number. Then run pipeline coverage analysis. If your enterprise sales pipeline coverage is lower than 300 percent, you likely won't make your target. If your pipeline coverage is 300-plus percent, but half of it sits in two massive deals, you likely won't make your target—unless a tailwind brings deals out of nowhere. No one is ready to bet on a lucky number or on Christmas miracles. Still, in response to the low pipeline coverage question, I keep hearing, "Fourth quarter is always big. Deals will come. That always happens." A bet on double zero is not a sustainable business model. Be practical.

Once sales coverage and sales dynamics are clear, move to pipeline review. Do not complete this step any earlier in the process. Play the bad cop with your salespeople. Criticize the stage that each deal is in. Otherwise you'll hear only what people want you to hear. *We're in good shape. No need to worry. We'll be fine by the end of the fiscal quarter or year.*

Problem verification is also vital when starting a new business or building a new product or service. Without deep, unbiased, numbers-based analysis, you can't say if the market opportunity for your venture is there (or not) or if the time to market has come (or it hasn't). Starting up is awesome. It generates a lot of excitement and optimism. And that's the moment when all terrible advice sounds like it was written in stone.

"Do what you love, and the money will come!" "Quit talking and start doing!" "Fail big, fail forward, fail fast!" These motivational quotes seem like a business plan that has worked for all those people who post Lambo and private jet photos. These quotes can excite fresh graduates who wasted five years at a university and have no clue what to do for a living. But thirty-year-old-plus professionals already in business? We should know better. "Doing what you love" does not necessarily make you earn more or build a successful business. Unless you have a brilliant idea, deep subject expertise, and unique capabilities to turn what you love into a cash cow, ignore those motivational posts on social media.

Success is not about passion but about problem-solving. What's the problem? For aspiring entrepreneurs, you need to verify that there is a problem in the first place. In either case, you'll ride the same line of questioning.

- How can you be sure that there is a problem?

- What can be done to fix it?

- What will it look like when the problem is fixed?

- Who will be involved in solving the problem?

- How do you make sure that the problem doesn't come back?

Once you're precise, you'll find your black-and-white answer. Identifying problems and finding the right solutions are vital to success. Both. Not one or the other. If you apply the wrong solution, even if you've identified the problem, you'll waste everyone's time.

In Kaspersky's META case, we ran into pretty much every problem you could imagine. That happens when you run five lines of business in sixty countries. Multiple problems can have multiple root causes. If a company is underperforming, maybe the product is bad. If the product is good, perhaps the marketing isn't relevant. Or maybe your key contributors have inadequate skills. You've got to find it out.

In a situation where performance doesn't match expectations, it's easy to turn leadership into seagull management. Too many leaders do this. I don't really believe in seagull management as a quick-win vehicle—fly in, scream during meetings, shit on everyone, fly away. Repeat as needed. I just don't think that coming in, hiring, firing, applying pressure, and asking marketing to be more aggressive will make a huge difference. It creates an illusion of hands-on management, but in reality it doesn't bring leadership greater influence over outcomes. In reality, seagull management makes everything worse. It kills morale. No progress is made. It's all shortcuts—quick fixes that might make you look like a superhero for a couple of months or pump up your self-confidence in what you're doing. You'd get some praise for easing the pressure on you from the upper management, investors, or business partners, but it won't get you any closer to identifying and fixing the real problems.

In Kaspersky's case, every problem had one shared root cause—an outdated business model that relied heavily on product excellence and a bunch of stakeholders stuck in the past. How we positioned our product, how we presented it to the market, how we pushed it through our sales channels, how we managed the sales and marketing teams—all of it was "the way it had always been done." The sales team was stuck in an endless beauty contest, entertaining customers and partners in the hope that they would buy. It wasn't working. We needed to make sure that every action had a purpose. If an average sales cycle was three months, we needed to know what was on the road map at every stage. What happens in week one, week two, week three, week four, and so on? Understand what's broken and how to fix it and what risks and opportunities exist on the other side. That was key to our success and will be to yours.

Tip 2: Always Have a Plan

Drawing up a good plan doesn't require much time, effort, or money. Actually, you don't need Big Four consultants to help you build a plan. It's a lot simpler than people think. Be practical. A lot of people overengineer and get stuck in the planning stage forever. Improving a plan to perfection never leads to executing it. Of course, neither does making zero plans. Find the balance. After all, preparation is half of victory. Your plans need to be properly precalculated based not on what you as the leader thinks but on facts. Data. Relevant information. Always ask for the expertise of people around you. *Everyone* has blind spots. You may miss something, which is fine if someone else is there to catch it. That's why I didn't build the plan myself or do so quickly. We spent three weeks on the same thing. When complete, I sold the plan to all three managing directors, my direct reports. I wanted them to understand that this plan was doable, that they were able to execute it. The last thing I wanted was to use my energy to convince everybody to make a kamikaze jump. I wanted everyone to buy into the plan for themselves and for the right reasons.

When you're asked to design a solid business plan, you likely imagine an almost scientific one-hundred-page document overloaded with numbers and formulas. It's hard to read and impossible to understand. In fact, a solid plan is much simpler that most people think. Here's how to go about it:

1. Make current state-of-business conclusions based on the assessment you accomplished (as described in the problem verification questions in tip one).

2. Set your targets and business objectives. Start from a high level and get granular after. Be clear about the monetary value of the desired incremental progress.

3. Identify growth areas among your product portfolio, customer segments, territories, new and renewal business segments, and employees.

4. Distribute sales targets based on opportunity proxies: current market share, YoY growth rate, sales coverage (number of customers or partners per account manager), revenue per head, existing renewal base, and external sales drivers, all specific for each territory.

5. Once distributed, ask how these targets and goals will be achieved. Each must be represented by KPIs and metrics that drive revenue.

6. Use a BI tool to generate the reports and dashboards you need to have a 360-degree view into sales and your business processes.

For example, let's say that you target 10 percent YoY top-line (revenue) growth and 5 percent bottom-line (profit) growth during the next fiscal year. You'll achieve this growth only through an incremental process. Apply your historical seasonality and distribute necessary increments across your twelve months. Review YoY growth rates every month. Correct any that are too high or too low. Remember, be realistic. Run revenue decomposition down to the KPIs and metrics. Set specific goals for each metric. Each goal must be aligned to the relevant sales and marketing activities that drive revenue. Use only actionable metrics in planning and execution. If your growth area is new customer acquisition, you need to set your goals around new revenue and the number of new customers. Align realistic sales and marketing activities, campaigns, and sales incentives to drive this metric.

I've used the word *realistic* several times because it's so important. Take a moment now to think about the most recent plan or process you've built. Challenge yourself. Is it realistic? Are you confident and comfortable executing it? Do you know what each person is supposed to do at every step? Person, not a role. Do you believe your people can act on it? Are they confident? It doesn't matter what you're doing in your business right now—starting your business, changing your business model, launching a new line of business, releasing a new product, or expanding to new territory. In whatever business activity you do, always start with reality. Is it executable? Are your people capable of doing it without micromanagement? Planning for the sake of planning doesn't work. I'm talking about building a plan that's actionable, realistic, and executable with existing resources. This is the easiest way to avoid micromanagement and to increase efficiency.

Once I'd shown the plan to my direct reports, it was their job to get their teams' buy-in. It was a tough sell at first. We were telling people what their revenue projections should be. These people were used to the occasional win, not daily excellence. Nor did they understand the big picture of the opportunity ahead at Kaspersky. Nobody ever explained it to them. I made sure to remind my managing directors of what their people needed to know—our goal and how we would get there, the challenges we would face, how we would overcome them, my personal commitment to their success, what my leadership role meant to their daily operations, and the role of each and every team member. If there was something we wanted to accomplish, we made a step-by-step, click-by-click process for it. Then we got clear on which roles did which activities. We designed every process to be foolproof.

Have realistic expectations from your team. If you have skilled people, you can visualize the entire process as you build it. Visualize every process and find metrics that represent your business model. For example, if you are looking to expand to a new territory, customer acquisition and transaction frequency are likely the right metrics for you. If you are looking to monetize an existing customer base, you'd better focus on lifetime value, retention rate, and average order value.

Sometimes you don't have all the right people on your team yet. When you build a new business or a new process in your existing business, you realize who you've got to hire. Poor process planning squashes a lot of young entrepreneurs or managers who are early in their journey. They don't visualize the process because they're in such a rush to start doing it.

Once you've built a plan and know who is doing what, you can drive your operations according to that plan without micromanaging anyone. I've often heard employees complain that I'm obsessed with numbers.

"You always say that business conversation that is not numbers based is pointless. Do you solve mathematical equations in your sleep?" one middle manager joked.

"I'm not a numbers guy," I said. "I'm creating a framework that lets you be successful. I'm studying the metrics so that I can understand exactly what's going on—the red flags and the opportunities. This way I don't have to micromanage you, and you don't have to micromanage your teams."

There is a great deal of confusion around micromanagement and hands-on management. Micromanagement is awful and extremely insufficient. It rarely brings expected results but ruins team morale, motivation, and confidence. Hands-on management is all about creating a framework and a business culture that leaves no room for micromanagement.

Hands-on leaders track their business processes with relevant KPIs and metrics that are empowered by analytics. They reach the alignment with their teams on how they read the data, how they communicate, and how they report back if needed. This sets managers in a position where they don't have to intervene in any ongoing processes because they understand what's going on there in real time. Having the luxury to put things in perspective through the data, leaders are able to see both the red flags and the opportunities and are able to address them with the team at the right time to accelerate performance.

Tip 3: Don't Cut Corners, Seek Quick Wins, or Take Shortcuts

While building and executing the business plan, I knew I couldn't cut corners. If I did, my team would as well. You can make any bad situation worse with incentives for unproductive behavior. You might hit a sales target this month or this quarter. But cutting corners can trash you next fiscal period. How? You'll end up overstretching your supply, shipping more products than your sales channel can digest, burning your marketing budget, or giving high discounts that set new price expectations you won't be able to meet again. Evergreen discounts aren't discounts. It's called price erosion.

Hiring is an area in which shortcuts happen even more often than in sales. If you've just stepped into a management role, you may decide to fill all open positions as soon as possible, prioritizing the speed of filling positions over hiring the highest-quality candidates. This creates a short-term feeling that your organization has been improved so that you are better positioned to achieve your goals and demonstrate progress to upper management. However, in the majority of cases, six months later you see struggling performance and internal conflicts because those new hires cannot cooperate.

Gaining quick wins can improve morale within the team. Achieve wins with excellence, and you have every reason to celebrate. The problem is, a lot of people *only* go after quick wins, and for the wrong reasons. They pump up an already filled balloon just to say that they put more air inside. They don't understand that it's going to explode next quarter or next fiscal year. You're not going to get a big win mid- or long term if you try to find a shortcut.

We tend to focus on what we like to do and ignore what we hate to deal with. In a situation where existing problems require a permanent fix, people often camouflage it or move their focus to potential quick wins. This is not productive long term. The problem won't vanish by itself. It doesn't matter how well you hide it.

A good example is when sales decline and sales teams face pressure from management. Everyone wants to hit their targets. What happens in most cases? To demonstrate progress, sales teams stuff sales channels with even more inventory or pull renewal contracts from next year. This makes it look like several account managers have achieved their sales targets, but it causes existing problems to worsen.

Another common situation is when customers churn due to decreased product or service quality and increasing irrelevance to the market. Most sales managers decide to address it with more aggressive discounts to show quick top-line revenue improvement.

This creates a ticking bomb that will explode during the next fiscal period and send the business into jeopardy. What should they do instead? Think long term. Build the capability to verify the root cause of the problem on the go and apply a fix that addresses both short- and long-term performance.

Evolution takes time. In nature, it takes millions of years. Don't expect instant transformations, but *do* expect progress over time. In organizations, the evolution of every person is a must if you want to be successful—evolution of your business model, business processes, culture, and, most importantly, mind-set. *We don't cut corners here. We do it right the first time.* You either evolve or you say goodbye.

Tip 4: Share the Big Picture

Make sure that every stakeholder one level down from you and one level above you understands the big picture, their role in the transformation, what good progress looks like, and what success looks like after that. Clarity helps you sell your plan to your team. If the people executing the plan believe in it, you'll be set up for execution excellence. At Kaspersky, I explained our position as reflected by a dozen fundamental sales performance metrics that were precise and easily measured. The crucial part of any transformational process's communication strategy is to guide your stakeholders from the current state of business and "why we're doing this transformation" to the ultimate goal of the transformation. You don't want to imply that you're going to fire low performers and terminate partners who have been with you for ten years. Rather, you want to communicate that business has been declining, 20 percent of top-line revenue has been lost, and more than 30 percent of the bottom line is gone. Talk about the reasons. The operational model doesn't fit the market anymore.

It's ten years out of date. The product portfolio is missing a lot from the competition, so the only way to succeed is to build an execution excellence framework (e.g., sell more with what you've got). The market is growing 11 percent YoY, so you see a great opportunity. You're wanting to double the business in four years, so here is how much you expect to grow every year. To seize this opportunity, you need to reshape the business model. You're going to offer a fair and transparent RFP to existing and potential distributors and pick those who fit and commit to helping to double revenue in four years. You're going to reshape the sales team to fill competence gaps. You're going to hire champions who will raise the bar and set new quality standards. This will be supported by extensive training and education for the existing team. There will be some layoffs, but those will be based on the new business model, not on arbitrary management calls. Once the transformation has been accomplished, everyone can expect to become the most consistently performing and profitable territory for the company.

This sounds a lot different from threats, right? However, both were explained exactly the same. At most companies, large and small, a lot of things have a question mark. Some people call it corporate bullshit. *Does it make sense to do this? Does it make sense to focus on that instead? What do we prioritize?* That wasn't the case for my team, as my communication strategy was an integral part of the execution plan. Everybody knew what to do, how to do it, when to do it, and why. As a result, we accomplished everything on our road map by year's end.

Everybody works for money and fulfillment. This is what makes people do their jobs, perform their duties, and hit their targets. But when it comes to something bigger than day-to-day tasks, you need to share the big picture to boost the team's motivation. Nobody is motivated by money or pride alone. People are inspired most when the plan makes sense. When they're clear about their role in the plan. When what they do makes sense. When they're clear about their contribution to the big picture. That's when you get buy-in, and your people commit indisputably to align with your plan—I repeat, *indisputably*. That means a red line. You invested a lot of time and effort into communicating the plan, your expectations, and all the *why* and *how* details. *The plan is ready. People have bought in. No disputes, no changes. I'm with you as your leader. If you're not in all the way, you're out. Now let's go execute.* Anything less, and you've got a kindergarten.

When your team understands the big picture, you will see commitment and dedication to the goal. That means a major improvement in productivity. But motivation doesn't last forever. You must repeatedly communicate what you want to achieve together and why you are confident that the plan is realistic. For example, guide your team through the market opportunities or your new killer product release. How are you going to achieve it while continuing your current progress? Your communication must all be data based. Make sure that everyone reads the data the same way you do. Leave no room for misunderstanding or misinterpretation. Create your own communication template for your plan so that your team and external stakeholders can read it as a story. I recommend keeping the same template (email, presentation, etc.) unchanged for at least three months. When the look and feel are the same, it's easier for people to navigate.

Make sure you communicate any course corrections openly. The last thing you need is for your team to see that inspiring plans don't match current performance and that their leader has nothing to say about it. If employees think that their leader has no plan B, morale is demolished. Performance suffers. Build confidence in your team's confidence in you and in your plan. Be transparent. Communicate frequently. It's not that complicated when you think of it as a process.

Tip 5: Leverage the Worst-Case Scenario

While my plan at Kaspersky was based on aspirations like doubling top-line revenue, we budgeted P&L and executed based on worst-case scenarios only. We created a road map *assuming* that the plan wouldn't click right away. Then we developed a clear path through every obstacle I could think of to make the positive outcomes as sustainable as possible.

Why plan for the worst-case scenario? Because whether everything goes wrong or not, you'll do much better than if you plan for smooth sailing. It's psychology and how P&L works. I knew that reality would most likely be better than the worst-case scenario. This meant my team would see more progress than what I'd mapped out for them. *This is working! We're starting to deliver! I can achieve my target as we have planned it. We're moving to the winners' bench!*

We built the plan to include logistical complications like me being in Los Angeles while the team was based in the Middle East and Africa. I set up an instant messaging channel to ensure that we would make the right decisions when something went wrong. And something *always* goes wrong. For us, that was distribution optimization in the first place. A lot of our partners weren't happy about the RFP. Most of the worst-case scenarios we mentioned in the road map materialized. We faced all possible pushbacks, from RFP sabotage to aggressive escalations to the board of directors with "Look at what your people are doing" complaints. Our massive hiring and firing decisions also caused complications. Headcount balance was always on the brink. Hiring and firing were never in balance. The risk of overspending when a sales ramp-up was not guaranteed was more than tangible. I've seen this risk materialize many times before. Corporations throw what seems like endless money at a problem, and business owners run their operations on out-of-pocket money. It's not smart to hope that one day things will turn around. It's better to plan to succeed in spite of the worst-case scenario. The steps to plan for the worst-case scenario are rather simple:

1. Budget for expenses as if you're going to achieve only 90 percent of your top-line revenue target.

2. Assess what is critical for success. Separate all the expenses into must-have and nice to have. In the majority of cases, you will be surprised by how big the room is for reducing expenses. Challenge all internal expenses like events, travel, and entertainment. Are these really necessary? Lock 20 percent of your marketing budget, and don't release it until the moment you are projecting solid top-line overachievement. Stay lean until then.

3. Assume gaps in sales coverage. If people leave and you are slow to hire, or if new hires are slow to ramp up sales, you need to have a clear understanding of how you will cover customers and partners with existing resources. Understand this on the level of specific people—who will need to run the extra mile? You may need to invest in incentives so that salespeople are willing to do this.

4. Have a plan B ready, both for the sales recovery and for the expenses optimization. Identify the activities that have minimal impact on the short-term sales results and get ready to cut them if things go wrong.

There is a scenario on how to reload the mind-set of the team that underperforms for a long time. What happens when the sales and marketing team is missing their targets quarter after quarter? They become losers at their jobs. And many of them are OK with it. *It's fine that I am not achieving my sales target.*

It's OK that we are losing market position. I'm still receiving part of my bonus package and a base salary. I have benefits. And I live nearby. They start to feel like they can't achieve their targets no matter what they do. Partners don't take them seriously. Customers give them terrible feedback. Negativity builds up, and you start to disrespect yourself. You don't believe you can improve.

If you want to make these people believe in themselves again, that they could work hard and smart and make good things happen, you need to help them. Success in sales is impossible without confidence. Avoid bullshit motivation like "We're going to make it! Just believe in yourself" or pressure like "It's your wake-up call!," neither of which would make any difference. People need proper guidance. Once they are clear about *how* they are going to make it, what obstacles they'll face, and what the plan B is to overcome them, things will change.

Because we planned for the worst-case scenario, I never lost control of the situation. Even with constant fires and explosions all over, I kept my confidence in front of the team. Of course, you can't be prepared 100 percent of the time. Nothing you do is guaranteed to work. But I knew that the simple plan we built, the metrics we influenced, and the worst-case scenarios we planned for would secure our progress.

Leverage your worst-case scenario to build your team's confidence. Any plan based on massive aspirations and high expectations is flawed. Prepare for the worst-case scenario. Because overachievement based on execution excellence inspires.

Tip 6: Set House Rules

Apart from the process excellence you want to achieve, make sure that you set and commit to clear house rules. House rules serve the big plan's purpose and stimulate a winning culture. To accomplish the business model transformation successfully, we set several of them.

Rule one: no excuses. No one was allowed to make any excuse, whatever happened. If your revenue forecast is falling down on a Thursday over month three of the fiscal quarter, no excuses are allowed. You need to come up with a clear plan for how you will try to compensate for your own underperformance. What's your plan B, plan C, and plan D? Not the bullshit "I will do my best" or "We will work very hard."

Rule two: no bullshit. If there is a problem, talk about it openly. A BI infrastructure and numbers-based decision-making pointed out the problem anyway, with no punishment for missed shots. Just come up with a recovery plan. Bullshit must be eliminated, as it has very little correlation with actual business outcomes.

Rule three: personal accountability. Every person was responsible for their and their team's deliverables. That is not the case for the vast majority of even successful corporations. In today's corporate world, the understanding of *we* is a form of CYA. I hear managers say a lot more *we* in bad situations and a lot more *I* when it's time to report a success story.

Rule four: playing politics is outlawed. Playing politics at work happens when people try to compensate for their own performance gaps. My vision for corporate politicians is pretty simple. If a person has time to talk about people behind their backs, corrupt the facts, and manipulate people, they have little time left for work—the work we pay for. If any senior manager, whether that be middle management, the vice president, or the chief executive officer (CEO), allows themselves to be manipulated by corporate politicians, that's a big red flag for business consistency. It means they have no basis for unbiased decision-making. In most cases, the absence of hands-on management is the reason why politics exists.

Rule five: no mediocrity. That word is not in our dictionary anymore. It is not allowed. If you want to make a difference and accomplish something remarkable, make sure you leave no room for mediocre people, processes, and attitudes.

What happens if you fail to follow one or more of these rules? The worst failure of all. Set the rules of your house wisely and always enforce them. Your success depends on it.

Tip 7: Be the Leader You Wish You Had

Once you've identified any problems, found proper ways around them, built a plan, achieved buy-in, began execution and numbers-based performance management, and set the rules, add the vital element—the leader you wished you had.

Let your people see you as their leader with your smart, detailed, well-justified, precalculated, and realistic plan and execution excellence. It's not bullshit, illusion, or fantasy. You're setting the example. Executing. Focusing on goal achievement. You have no choice. Because in the middle of your plan, you will lose your temper, your passion, and your commitment. It happens much too often. You will feel ups and downs. It will be like golf, a constant chain of excitement and disappointment. That's business. That's life. What will you do? Become a seagull manager? That's a problem. You will ruin everything. Your credibility. Your reputation.

To fix those will be difficult. Managers tell people what to do while leaders inspire them to do it.

Throughout the execution, you will be pressed, banished, and trolled. Call these haters if you like. Surprisingly, those haters can be colleagues or even key stakeholders who don't deliver or perform. Your success makes them feel bad. Sharp words make nobodies feel better about themselves. Decide now that you will not give a shit about what they say. Dismiss *all* criticism that is not based on data. No data, no conversation. Feedback and constructive criticism from credible people is always helpful. To be criticized by the people who care, are capable, and can share their expertise is great. I love to be criticized. Seriously. People are people. We can't be perfect all the time. For all others who criticize for the sake of criticizing or looking better while criticizing, ignore them and keep doing your thing—stick to the plan.

Better yet, prove them wrong. When I became Kaspersky's META managing director, both management and employees had already witnessed broken promises. That's the story of just about every organization on this earth. The people at the top overpromise and underdeliver. That's why I led differently. When people saw that I started to deliver what I'd committed to, they executed the plan with such excellence that they didn't need me there to make their region a shining example for the entire industry.

At the end of the day, isn't that what everybody wants? To motivate others to achieve big. To lead without apology. To make your company a great place to work. To do such a good job that your presence is felt long after you're gone. That can be your reality. Follow the path I've laid out for you with these seven tips, and I expect it will.

CHAPTER

Dream Big: Terrible Advice about Ambition (And What to Do Instead)

Remember the last time someone said to dream big? Did they mention that this big dream needs to be realistic? Probably not. We live in a society obsessed with *bigger*. Bigger houses, bigger plates, bigger dreams. Especially bigger dreams. When was the last time you heard someone say that they just wanted to be content? Or that they wanted to work hard and earn more responsibilities?

Public success stories mislead many talented people who misunderstand the concept of dreaming big. What's rarely talked about? Picking a realistic dream, anticipating constant failure on the way to achieving it, and appreciating the long, nonlinear journey.

We all know people in our circle who often talk a big game, but few of them ever deliver. They've got big dreams with a big ego to match, but they don't follow through. Dreaming big might be great for setting milestones, achieving them, and moving forward. Or it might be an excuse for why you're not having success. Like any task, a big dream requires a great deal of work and patience. People rarely commit to a realistic goal. Most people *only* dream about big things instead of working hard to get them.

In response to the hollow popular advice to dream bigger comes minimalism in living and saving. Is it better to feel satisfied with what you have than to always hunger for more, more, more? All people are different. Ambitions are different. Minimalism is another extreme—the opposite of dreaming big. The truth is somewhere in the middle.

One Day You'll Be CEO:
An Overnight Success Story That Took Ten Years

It was 2004. San Francisco. The global Seagate Technology sales meeting. Seagate, the famous company that builds the hard disk drives for nearly half of all computers and servers on the planet, flew remote salespeople like me out to meet up with the rest of the sales team twice a year. This meeting's crowning event was a keynote speech from a special guest. I don't think anybody was as awestruck as I was by the speaker.

Taking the stage in one hour was General Thomas Stafford, the world-renowned astronaut and commander of the 1975 Apollo-Soyuz Test Project flight (the first joint US-Soviet space mission). A Russian spaceship called *Soyuz* (which means "union") and America's *Apollo* connected in orbit for a few weeks. This flight was the predecessor of the International Space Station. It was also the turning point of the Cold War, in my opinion.

While some of my Seagate colleagues had never even heard of General Stafford, I'd learned about him and other astronauts my entire life. My parents spent most of their careers in the Russian space program. As a kid, I met astronauts in my dad's office and at school when they visited to speak. Meanwhile, students my age around the world learned that the Soviet Union was an empire of evil. That was true of the communist regime but not true of the people who had no choice but to survive in the Soviet Union.

I believe that General Stafford changed how people felt about Russia and her people. To see this remarkable man in person was, for me, priceless. I walked up to the head of the event team and asked for an introduction.

"There's a gentleman who wants to talk to you," the organizer said to General Stafford. "Maxim Frolov, sales manager from Moscow."

The general looked at me, stood, and shook my hand. Then he greeted me in Russian.

"What a pleasant surprise to come to an event like this and see a guy like you," the general said. "I love Russia. I love Russian people. I'm still very close friends with Alexei Leonov." Leonov was the *Soyuz* commander. "I visit Russia often, even at my age. My wife and I adopted two Russian boys."

"Amazing," I said. "I graduated from Military Space Engineering Academy as a space communication engineer. Every Russian who worked for the space program knows who you are. Hell, every *Russian* my age knows who you are."

He laughed.

The general and I sat together, chatting in private for some forty minutes. We talked about the space programs in both America and Russia. We never touched on business.

General Stafford shared wise advice during his speech. At nearly seventy-five, he carried himself with the energy of a young man. He spoke with confidence. One moment in particular stood out to me.

"I'm so happy that this business is going global. I'm so happy to see Russians in this room who work for a global American company," the general said. "Maxim." He gestured to me, and the whole room turned to look. "I wish for you to one day become CEO of a company like this."

Can you imagine me, a little guy, a big nobody, hearing *that*? And knowing everyone else heard it, too? I felt uncomfortable. Shy. Humble. I got goose bumps. No one outside my sales team in Europe knew I existed before that moment. I'd handled hundreds of millions in sales in my two years at Seagate, but I was one of a thousand people working for the company.

For a simple Russian guy like me, there wasn't a career path to CEO of a global company. Until General Stafford's speech, that hadn't mattered to me. To work at an American company and be responsible for hundreds of millions of dollars in annual revenue in as big a market as Russia was enough for me. When you have no formal schedule and no boss hovering over you, it's easy to not take your career seriously. I thought I was happy in sales. For the first time, I questioned that. For months after General Stafford called me out, I put myself back in that moment again and again.

Now what? I thought. *Why not become a CEO of a global company one day, as I was advised? What's stopping me?* A scarier question popped up next. *How am I going to do that?* I didn't personally know anyone who had made it to C-level. When you work for a corporation, C-level people look like superhumans with abilities not available to others. They speak a business language you hardly understand and make decisions of enormous scale. When I compared apples to apples, I raised even more questions. *Am I good enough to target the C-level career path? Is there a road map for the remote-based salesperson to become CEO? If there is, is it even realistic?*

I lacked experience. I might have been a hero in my neighborhood, but I was a nobody beyond my arena. I needed to know the vital processes of my counterparts' work and what made them successful. I wanted to learn how to run sales operations efficiently and how to make newly hired sales teams deliver on target. I stepped up to run planning, forecasting, and partner program designing as well as to do new hire onboarding and coaching.

I set clear expectations for myself. Number one, understand the big picture. I outlined every business process from A to Z, both my own and those in other departments. When you understand every aspect of the supply chain, the sales and marketing process, deliverables, risks, and opportunities, you see where you could add value. That's what I did, even proposing these to management. I asked for more responsibility and a greater workload, not a promotion or a raise. Make me the team leader without the actual promotion. Give me opportunities to learn at work. That's how to earn a promotion—and move one role closer to CEO.

I took on the extra responsibility for sales forecasting and coaching new hires. I'd troubleshoot problems with the marketing team. By 2007, my efforts started to pay off—I was awarded Seagate Sales Manager of the Year. Not just in Russia. In the world. And for good reason: We were the indisputable market-share champion—Seagate now owned 60 percent of the hard disk drive market in Russia. Second place wasn't even close. Moreover, sales revenue had almost doubled during my five years of service. This increased my visibility and credibility at the top management level. It was a milestone I had to monetize.

Now am I close to becoming a director? I asked myself. *What's next?* I didn't wait for the promotion because that's not what you get when you're just a solid salesperson. I kept earning the new job by coming up with a plan to target new lines of business. One new line was retail—brick-and-mortar retail businesses for hard disk drives. Today retail is losing ground globally under the pressure of ecommerce. Back in 2007, technology retail rocked, and I wanted to be part of it. Seagate had never done this before in Russia while our key competitors had. I developed a realistic, executable plan, found the right people to hire, and convinced top management to give me a shot.

"We're late to the retail market in Russia, and all our competitors are dominating the low-price segment," I explained to management. "I think we should position Seagate as a premium brand and start to sell premium lines only. We'll never win in the overcrowded retail space if we're competing on price only. Let's target customers who can afford to pay more. Let's offer premium products with a solid marketing story." This turned out to be a fast-growing new line of business for us.

At the end of 2007, I was promoted to a director position. Phase one of my road map was accomplished! My first year as director brought unbelievable excitement. Betting on the premium product line worked. The new line of business ramped up much faster than we expected. The team I'd built was killing the market. For the next three years, we continued to run the show in Russia.

That consistent achievement brought me more frustration than I expected. Despite the hard work, everything seemed nice and easy. There was very little failure there, so it was hard to tell if I was that good or if organic market growth had helped me. I was making great money, hitting my sales targets, delivering good margins, and exceeding company expectations. But what was next? I doubted that I had the management skill set needed to level up. Gaining them at my current role wasn't realistic due to the nature of our business model. It was simple and easy to execute. Not the place to get practical experience running P&L from A to Z. The question was, where would I acquire it? A side hustle seemed like the only way to gain the experience I needed to become a CEO of a global company one day.

I experimented with side businesses while I worked for Seagate. One of them is worth mentioning because it was a rich source of lessons. My friend and I invested our savings into a digital marketing agency start-up. At that time, it looked like an easy money business where in-depth subject knowledge wasn't required. Everyone wants to have a cool website and get it promoted, right? And two great sales guys can sell anything, we thought. Order fulfillment will just happen one way or another. This is a very naive and impractical thought process if you want your side hustle to become a second source of income.

We decided to build a hybrid team. We hired website developers, mobile marketing experts, a graphic designer, and video producers. But our sales experience met a steep project management learning curve—we had zero know-how for managing diverse remote teams. The idea of starting our own company and making easy money was nicer than the reality. Firefighting wasn't a problem, as we were used to dealing with it at our current jobs. The problem was strategy and the scalability of the model. The market was so overcrowded, we barely beat competitive offerings. Margin was a big question mark, as it was extremely difficult to understand net cost during contract negotiation. We tried to improve by focusing on more value-add projects like 3D modeling and virtual tours for real estate. But net earnings didn't show a big difference because expenses grew faster than revenue. After just one year of operation, we had to shut down the agency. Our remote teams couldn't fulfill the orders we were bringing in. We failed to recruit and manage them proactively. We walked away with net zero return on the money we invested. We got a great lesson about the production process, quality, and agility in the overall success of a venture. The learning curve was amazing, as we had dedicated all our free time to that project for an entire year. The failure of our business strategy helped us develop execution excellence skills, as that was the only way to compensate for our missed shots. As we thought about our side-hustle failure, we realized that we had made another strategic error besides underestimating the skills required to run a technology business. We wanted quick money from this business. We didn't appreciate the importance of the journey required to earn it. This failure helped my friend and me make smarter career directions. I was able to navigate toward the C-level, and he built his own company, sold computer accessories under his own brand name, and has grown successfully for eight years and counting.

While successful and comfortable running business for Seagate in Russia, I could no longer be the biggest fish in a small pond. I had to confess to myself that my sales job wouldn't take me where I wanted to go—running international business at the C-level. It was unlikely that my knowledge of business in the former Soviet Union would be useful in a different business in another region. Seagate's business priorities were about to change rapidly. The cloud disrupted the tech industry. Apple, Google, Amazon, and Microsoft soon became Seagate's major customers, using our products to build their cloud storage infrastructure. This was the future for the hard disk drive business—not what I'd been doing successfully for almost a decade. Still, I was sad to leave. Seagate had a unique culture when I worked there. There was very little politics. In a company of fifty thousand employees, the sales and marketing organization was only a thousand people worldwide. We knew each other by name. We were all about efficiency, transparency, and impact.

What skills and experience do I need to be successful in this new world? I asked myself. *Where can I get them to move toward a C-level position?*

My answers came in the form of a job offer from a little-known American computer business called Microsoft. In 2012, I accepted an offer from Microsoft Russia to join them as a director for small and medium business.

Software sales models are much more complex than hardware original equipment manufacturing (OEM) and distribution ones. Software sales success requires different competence. Emotionally, it was tougher than I expected. Feeling like a superhero for ten years gave me a reputation that I worried I could no longer live up to. For my first six months at Microsoft, I felt extremely incompetent. I had to accept reality—*I'm not the big fish anymore. I'm surrounded by better professionals, and I need to absorb their knowledge quickly.* Every little step forward required huge effort. None of my past wins counted. I had to rebuild my track record from scratch, and I was missing a lot of skills. Critics easily offended me. I realized that I wasn't nearly as cool as I thought I was. It turned out that I had ego problems. Fortunately, Microsoft cured those. It felt like I'd never worked that hard. Understanding that you still have the capacity to work harder, learn complex things faster, and deliver results was awesome. That job was like MBA coursework in the field. My two years there made a big impact on who I am today. They suffocated my ego, took all my potential, and, through a painful workload, elevated my skill set to another level.

In 2014, when Kaspersky offered me the international managing director role to run the business in more than one hundred countries, I was ready for the responsibility and equipped to make a difference. And it was required from day one, as Latin America, the Middle East, Africa, and Central and Eastern Europe were experiencing a deep crisis. I was hired to fix it without breaking what was working on. Learning a new industry on the go—cybersecurity—and crisis-managing such widespread territories brought an amazing learning curve. I hadn't managed five different lines of business across different time zones before. This gave me the international experience I was missing and helped me develop executive leadership skills and strategic thinking that played a crucial role in success when I became CEO of my own technology start-up, Urban Innovation Group, one year later. I was finally ready to become a CEO who was capable of running a business in multiple time zones; managing diverse teams in Russia, Spain, and the United States; and managing C-level partnerships and investor relations. My "zoom-in" (identify and solve the problem) and "zoom-out" (correct the strategy) skill set got the ball rolling.

In today's social media-influenced world, it's easy to call yourself a CEO. Developing yourself into a C-level leader who generates results in any environment is a different story. The evolution of professional skills and mind-set is a journey rarely appreciated. Perhaps a complete road map of this journey will give you more value, fulfillment, and fun than accidental overnight success ever could.

Tip 1: Deploy Self-Awareness

Be realistic when choosing your big dream. Your dreams must correlate with your abilities. Big dreams and mediocre performance don't go well together. If you're an average performer who's not growing in your current job, deploy self-awareness and find the right path for yourself. Understanding who you are and what you're capable of is key to making your dream a reality.

Almost every day I meet people with a successful career in the technology industry who want more. They feel inspired to launch the next unicorn start-up, build a billion-dollar company, and make history. They're eager to earn more money than they have at their nine-to-five jobs. They often have the full support of people who want to join them, but in many cases, the results are unfortunate. Publicly available stats say so.

What goes wrong? Many excited future entrepreneurs often don't understand the full picture. When you have a nine-to-five job, you're part of the business machine. The machine works at its own pace.

Everybody at the company is a part of a complicated process. You might be a superior employee, and you might deliver excellent results year after year, but you likely don't know the details of what happens before or after your part of the process. Unless it was your job to do so, you don't know how to properly manage people and business processes outside your team, make corrections to a plan, launch or relaunch a product or service you sell, alter financial performance and investment priorities, or make the business as a whole successful. After your involvement in the business processes, which I'm sure you know how to do very well, you don't know what exact steps to take to start, run, and grow a successful business. These same challenges apply to people who want to get promoted in their company or move to another company or industry. Where you will go is very different from where you have been. As Marshall Goldsmith says, "What got you here won't get you there."

Many midlevel managers who I've watched leave their corporate positions to become CEOs of their own start-ups promptly fail—and go back to their old jobs. Many CEOs you come across on LinkedIn or any other social network are actually a one-person company who declare their intentions to make it rather than actually making it. I'm always surprised to see people who *didn't* realize superior results as employees start a business and expect success.

If you think you have what it takes to run your own company and then you fail, there's nothing wrong with that. I've said it several times before. Every person needs to experience self-employment and to try to make something on their own. Even if you fail, you can walk away with no regrets. The question is, how do you do it right so that you maximize your chances of success?

Many people underestimate an important factor for starting a business in the same field where they've worked as an employee—you need to be excellent at everything you did. You need to be a superstar who did a better job than anyone else in the company could to maximize your chance to succeed as an entrepreneur..

If you're a superstar, if you made a difference, if you became your own brand at your job because it was you who kept customers around (not the company itself), if you became a person everyone could rely on, then you have a great chance to succeed. Why? If customers used to choose you, not just the company you used to work for or the product you used to sell, there is a great chance that they would like to continue the relationship—with you personally.

By brand, I don't mean that everyone likes you. I mean, does everyone know who you are? Do they respect you? Does your reputation precede you? Do you have credibility with your direct reports, peers, and customers? Do you have leverage to make customers buy, or are you just holding a large collection of business cards? If you called a coworker tomorrow and said that you had a brilliant project to start, would they join you? If you had a conversation with a customer the day after you quit your job and said, "I'm on my own. If you remember how great my service was, I can promise you that it will be even better because I don't have the limitations of bureaucracy," and they say yes, they'll choose you now, you've got a brand.

Most of the lean one-person businesses succeed because the founder-owner made a huge difference for the customers they worked with at a previous job and was equipped to run the entire business process on their own. If your customers are happy to follow you in your new venture, you have a brand. If customers choose your company because they are first choosing you before you make the decision to say goodbye to your job, you may be in the right place to work for yourself and to do for customers exactly what you did while employed. If you don't see customers follow you the same day and make you a supplier of choice, then you might be fooling yourself.

Be realistic about your ability to execute your business model from A to Z on your own. Even if it's a fairly simple small business with one type of product, one sales channel, and one type of customer, there could be many areas where competence is required to take advantage of revenue growth opportunities. Do you have that going into the new business? You'd better.

Understand who you really are versus who you want to be or look like. If you're a successful manager in a big corporation, your track record backs you up more than you might think. If you're used to giving orders, hearing reports, correcting errors, and playing the game at the top, you face a harsh truth when you run your own small venture. You're now the person you were ten, twenty, even thirty years ago. You're a beginner who lacks experience again. You have different roles as a salesperson, a marketing manager, and maybe even as head of product development. You'll have to get involved in areas outside your comfort zone, areas delegated to other people back when you were an employee. Until you can afford to hire people, these responsibilities fall on you. There's a huge difference between telling people what to do and doing it yourself.

Similar scenarios unfold when managers who've built their careers for a decade within one company move to another industry. Many of them fail or experience massive mental and emotional pressure because they only know how to be successful within the environment of the company or industry they worked in the longest.

In the majority of cases, these skills are not easily adapted to fit the new reality. This is exactly what I experienced when I moved to Microsoft after ten years at Seagate.

Many high-performing people confuse past success with future ability. If you've been a solid performer and you start to dream big about becoming a C-level executive or a business owner, you need to assess how you achieved your current success and how scalable it is for the new venture. Did you hack the business? Transform the business? Rewrite the business process? Or did you just do what was expected in a growing market with a full pipeline of highly motivated customers who wanted to buy an awesome product that other people in your company developed? In other words, were you responsible for your success, or did you get a high-tide push? Answer honestly. This sets up the stage to outline your actual capabilities versus the capabilities needed for your next assignment, career move, or start-up.

Your big dream goals must match the actual skill set and business experience you have—not imagined skills and a track record you can't take credit for. A global economic recession is here. Self-awareness is vital to survival. When economic slowdowns force companies to reconsider their P&L, cut costs, and lay off employees, those who think they're all that are always the first to go. The market always has the final decision.

Recently, an old friend asked me for career advice. After more than twenty years in business, he'd reached the role of country manager for an international company, a job with a small scope of executive work. He led a team of four people. He did most of the sales and key account management himself. When his employer faced performance challenges on a global scale, he was the first to be terminated. One of his team members could do the same work for less money.

My friend felt stuck. He was frustrated with the job search. No one considered him for another country manager or sales director role. Why? I explained that his competence was key account management, an individual contributor role. He had no experience managing P&L or full-scale sales operations, even though those were part of his past or his last job description. I asked him to outline what top management abilities he was able to bring to the next job. He found out that all of them were around how to make a target within a few key accounts. I told him there was nothing wrong with being a key account manager at a bigger company, as that matched his actual skill set and experience. I advised him to be realistic; the job titles he'd had did not reflect the skill set required to be a successful GM. I'm happy to say that he followed my advice and joined one of the largest technology holdings in Europe. Don't let titles confuse you. Understand what skills you have, what value those give your new venture or employment opportunity, and where you can make a difference.

Self-awareness is an ongoing process. You can't do it just once. You have to run this exercise on a regular basis. Get used to writing down your core competencies and shortcomings in your current role. Admit that nobody is perfect, so knowing what you are good at and bad at is crucial. Count your strengths and figure out how to amplify them to close any gaps you have. Once you understand what you are not good at in your current role, figure out how to fix it.

There is nothing wrong with feeling incompetent or lacking skills. That simply means you are at the beginning of the great journey toward your big dream. You may decide to delegate what you're bad at and focus on developing your strengths. But keep in mind that there are certain skills *every* leader must have. For example, you're unlikely to succeed if you bet everything on your entrepreneurial creativity while ignoring your complete lack of low-performance management experience and finance management skills. Find the right balance that works for you personally and for the tasks that you deal with every day.

Self-awareness has the biggest impact when it comes to deciding on your next career step. Thinking about taking a bigger role in another company or investing in your own venture? Make sure that you're aware of who you are and what you're already able to do in those new, tough situations ahead. You'll get them all, so it's all about facing the challenges properly armed.

As you start to measure who you are versus who you think you are (or who you want to be), self-awareness will become a habit. Don't try to look better than you are. Admit your mistakes. Focus on recovery and improvement, and you'll experience consistent growth. You'll set yourself on a clear path toward who you want to be and what life you want to live. A surprising side effect of self-awareness is awareness and appreciation of others—especially those who work for you. Sustained self-awareness will make you a better person and a more efficient business leader.

Tip 2: Pursue Realistic Dreams

Once you know what you want, you *can* create practical steps to get there. When people think they're pursuing a dream, they often act like all they have to do is talk a lot about it, visualize it, and let the dream materialize. Thoughts create reality, we're told. To some extent, that's true. But if you just *think* you'll be successful and do nothing about it, nothing happens.

Be practical. Do the work that your dream requires. Get clear on the road map you'll follow. If your target is to grow your salary from $100,000 to $200,000 a year, understand what skills or experience you're missing. Be realistic about the scenarios that you'll face and how you might correct course along the way. Working hard and dreaming big is not enough to level up. You have to have a realistic road map and execute it daily. How exactly are you going to do that?

If you're employed and want to develop your career, study realistic income data for the positions, locations, and industries you want to work in. Break down each part of the career into its core elements, including the following:

- Know the difference between industry trends and wages. Are they growing, stagnating, or declining? At what pace?

- How big is the difference in earnings between each level-up position on your career road map?

- What required competencies and practical experience do you have that add value and that employers are ready to pay for? No generic answers. All must be industry, geography, and company specific.

- Differentiate your self-assessment results from the requirements. How can you get the practical experience and competencies you're missing today? Outline realistic scenarios only.

It's also important for employees dreaming of escaping nine-to-five slavery and starting their own business to understand the facts. Because they're brutal. Most ex-employees who start a business won't earn anything close to what they made as an employee. If you are choosing the independence of entrepreneurship, get ready to have twice less income for the first couple of years of being your own boss, with a chance to accelerate after. This is how it works in the real world. In most cases, fresh entrepreneurs simply lack the required sales, marketing, business administration skills, and mind-set to stay lean and low key. I've noticed this reality in many industries, but especially in technology. There are so many brilliant success stories about young people from the third world developing a few lines of code, getting their start-up valued at $1 billion-plus, and making their dreams happen. Everything seems easy when you read the headlines because you're looking at the final destination and ignoring the fact that the journey very likely wasn't smooth. It's an absolutely different story when you're developing and marketing your technology product while also pitching venture capital funds to invest in you because you are running out of cash. Too many fresh founders are either not ready to sustain this heavy workload long enough to see returns, or they keep dreaming about success instead of executing.

When we launched Urban Innovation Group back in 2015, we were inspired by the uberization of technology. We had a great vision for the perfect smart city management platform, so we counted future success in the hundreds of millions of dollars. Our presumed business model was a hardware-agnostic platform that would convert any traditional parking facility into smart parking. The whole world would say, "Wow," and organic growth would happen globally. The beauties of our model were unparalleled scalability, low cost of customer acquisition, and low cost of maintenance. It was all true apart from the fact that the transportation and parking industries were heavily regulated and lived within exclusive long-term contracts only. In other words, our pay-as-you-go model didn't fit the market. We had to digest this harsh reality, understand that disruption of the consumption model was unlikely within the time frame we assumed, and reshape our go-to-market strategy. In a short period, we went from our rolling organic growth model to an enterprise sales model and altered our P&L and commitments to investors. We stuck with an uberization motto, but our goal was to reach positive cash flow, finance product development, and educate the industry to shorten time to market. This changed our financial expectations, but we did what we had to do to keep the start-up alive.

Later in my career, there were times when I quit full-time work to run several projects simultaneously. At that point, I could afford to do that. Also, my comfort level with having no income is likely higher than most people's. If you're considering starting a business, count on having no income. If you have no income, do you have cash saved to cover a year of living expenses? Or have you built passive income, like money in the stock market or rental properties? If you can live for a year with no income without changing your standard of living, it's safe to drop your existing career and go all in to start a business. You've still got to be realistic about your chances of success and failure. Set expectations about what this journey can give you—and what it can't.

The choice to go into business for yourself shouldn't be about money only. If you feel that you'd be happier working for yourself and sacrificing part of your income, go for it. That's a good choice. Do you want to run a side hustle in your free time while you keep a well-paying job? That's practical. Safe. Productive. A side hustle is the best personal development exercise in the world. I've done this many times throughout my career. A side hustle impacts your current job because you learn how to run a business, outsource tasks, and make every penny work like a dollar.

Before you give something up to go after your dream, make sure it's worth it. Make sure this is exactly what you want and are capable of seizing. If it is, *own it*. Let your approach be practical and specific. Have a plan B. Be willing to sacrifice a lot. And know that even if your dream is practical, you will forever be a work in progress. Nobody ever arrives. No dream is ever completely fulfilled. And that's all right. Progress is what makes us happy. Your journey may or may not be successful. It's up to you to gain experience, learn from ups and downs, and keep moving in the right direction.

Tip 3: Kill Your Ego

Before joining Microsoft, I understood that I had gaps in my competence. I didn't realize one of those gaps was my ego. Ego problems are easy to identify. If your first reaction when people criticize you is to get angry or offended, you've got an ego problem. My boss, the global vice president of small and medium businesses, gave me advice I'm grateful for to this day: "In your new complicated role, you'll feel incompetent most of the time. That's fine because you'll learn how to handle that. A constant hunger to learn more, to be capable of more, and to adjust to fast-changing realities is how you'll succeed."

I had that hunger. So I put my ego in check, took the blame, and decided to work harder to obtain the skills I needed to reach the next level. Checking your ego not only benefits you but also helps everyone around you. An ego on the loose has the opposite effect. Your ego is more dangerous if you manage people. Your ego impacts everyone. Be critical of yourself. If you're pointing fingers, you have an ego problem. If you refuse to read the warning signs, you have an ego problem. Decide now that anything that happens in your company, department, or team is your responsibility and yours alone. Always.

For every new chapter of your career, you must evolve. Constantly. Ego won't let you. Choose to let go. Become a student again. I treat every job as a learning experience. I learned how to accomplish more in a single hour than most people do in eight.

That's the kind of productivity you need if you want to achieve your dream, and you only get it when you kill your ego.

Tip 4: Alter Your Road Map

In most cases, the path you choose on your journey to the top won't necessarily be the path you finish on. Like me, your dream can shift as you learn more about your business, your industry, and yourself. The career path isn't linear anymore. How we answer the question "What do you do?" has changed drastically in the last ten years. Technology creates new jobs and terminates old ones. Migrations from one industry to another are accelerating. Remote work is no different from an office. Higher education is no longer required to get a job in the world's top tech companies. People are waiting to retire. Productivity is increasing across the board. There is no such thing as a wrong career move anymore. Employment now often takes the form of short-term projects. You work on one project for a business until it's done. Simple. Effective.

As you update your dreams, build new skills to fulfill them. This might mean you move from a C-level or management role to a start-up founder. To move from a high-paying global role where you have limits on what you can do to a lower-paying role in another company or another industry where you get decision-making freedom and where you can increase your skills makes a lot of sense. This might teach you zero tolerance for wasted budgets and constantly underdelivering employees who sell you their loyalty. Enrich your experience and skills with hands-on management where you are responsible for the entire company's results.

When you work for a corporation, resources seem endless. Mistakes don't matter. An army of people cover the process before and after you. Ten people might do the job of two because more looks better to some managers. But it makes any person who has ever run their own business sick. A dollar spent with no return is a waste of that dollar. Don't try to capitalize on your past wins by staying in your current job and hoping that the company does OK in the future. In the fast-changing world we live, this model is not sustainable. Be practical.

Today your career reflects what you want to do in life, not just how you get your paycheck. To get a lifetime's worth of experience through what you do has never been easier. The best way to experience the world is by working for a global business or by doing your own business globally. Thanks to technology, we all have the ability to do business outside our home continent.

Learn where the world is heading and what that means for you. What risks and opportunities are there in the business you have today? What's the most logical career step for you in the short term? What will you gain besides the rise in payout? What's the long-term impact of this move? When you follow realistic dreams with a clear road map leading to them, both your heart and your bank account will be full.

CHAPTER

You Deserve It: Terrible Advice about Entitlement (And What to Do Instead)

You know what I've noticed about our society? Entitlement rules most people's expectations. They believe they deserve better by default. Better life standards, better career, better achievements. Every inspirational quote that goes viral on social media is a variation of "You want it, so you should get it" or "Do what you love, and you won't need a job." Most people interpret these the wrong way.

The number of people confused about what they deserve boggles my mind. Scroll your Instagram feed, and it seems realistic to make a million dollars during the first year of self-employment. Most people who leave nine-to-five jobs to start a business earn less than they used to or would have been able to if they'd stayed employed. If you quit your job because you want to be happy, be your own boss, and do what you love, you're making the right choice. Most people start their entrepreneurship journey to make lots of money quickly, drive a Lambo, and fly in a private jet. A classic lose-lose scenario. Going all in to entrepreneurship requires not only balls but also a verified opportunity, a precalculated business plan that includes a plan B, the ability to execute under pressure, and realistic expectations. One such expectation is that you'll probably wear many hats. Up to 56 percent of all small businesses have four employees or fewer, including the owner.[15] In most cases, these businesses never grow beyond that size. If you're willing to work the jobs of several people, entrepreneurship may be right for you. As long as you have a plan and the ability to execute it twenty-four/seven.

You may have noticed that people who believe they deserve better circumstances, more income, and fewer troubles are the last to get what they want. Everyone has their expert opinion about other people's successes and failures, but rarely do they appreciate the hard work and sacrifices behind the scenes. Unfortunately, too many people confuse being an expert in a subject matter with actually being a businessman. In most cases, those who criticize a lot rarely deliver a lot. Keyboard warriorship is a fast-growing type of self-made expert who influences opinions and decisions at scale. This infection is spreading around both the corporate and the small business environment at a scary pace. Entitlement is a side effect. You don't *deserve* something because you want it. You don't deserve something because you think you are an expert who has a shiny résumé or education. The world doesn't owe you shit. Not your boss, not your customers, not anyone. If you're the kind of person who gets what you want through hard work, smart choices, and good old-fashioned sacrifice, you're the one who deserves it.

In the Soviet Union, Shit Doesn't Owe You: How I Started an Illegal Business at Eleven Years Old

The Soviet Union in the late 1980s was a strange place to live. The country was under communist rule, launching people into space, and building weapons to threaten the rest of the world.

---

[15] JPMorgan Chase, "Small Businesses Are an Anchor of the US Economy," Accessed April 23, 2020, https://www.jpmorganchase.com/corporate/institute/small-business-economic.htm.

Meanwhile, the economy was in terrible shape. Our few TV and radio channels were 100 percent government owned and 100 percent censored. Soviet ideology and propaganda were based on equality. Everyone was expected to be equal to everyone else. Get an average job. Get an average government-owned apartment, if you were lucky. Live an average life according to communist standards. That meant poverty. Equality in poverty was a fundamental principle. Soviet law regulated all wages. Everyone knew what you could earn based on your profession, your position, your years in service, and your location. It was easy to chart your life if you were happy living the Soviet way.

The economy was sick. Government-owned companies manufactured products no one was going to buy, but they created millions of jobs. This was normal. The average person with a university degree could earn the equivalent of fifteen to one hundred dollars per month. Sound miserable? You have no idea. We couldn't even use that money to buy anything from our wish lists because the Soviet economy filled the store shelves with rubbish. It's called a preplanned economic model. Every factory was producing stuff but had no idea who would buy it. It guaranteed jobs but brought no motivation for producing competitive products. All the good stuff was imported. The best goods were distributed to the elites or to specialty shops that accepted solid currency only (e.g., US dollars, British pounds, and French francs, not Soviet rubles). Sounds weird, doesn't it? It was. It was illegal to have foreign currency in the Soviet Union. Thousands of people ended up in jail for it. The only legal way to hold foreign currency was to get a permit to buy it at the bank. Such permits were granted only to travelers abroad and always came with an exit visa. Yes, you had to obtain a permit to leave the country first, then a permit to buy foreign currency. Once you got back, if you still had foreign currency, you were allowed to spend it in those specialty shops full of imported goods.

While being equal in poverty was the Soviet standard, trendy clothing was the only way for kids like me to feel cool and to differentiate ourselves. My parents, who both worked for the space program, could not afford the stuff I wanted. Their salaries barely covered the essentials like bread and milk. In the Soviet Union, a pair of Adidas or Nike sneakers cost a month of household income and was hard to find. Needless to say, when I asked my mother for a pair for my birthday, I got a resounding no. I understood that we weren't able to afford them. But I wanted them badly, and if my parents couldn't buy them for me, I would find a way to buy them myself.

My first business idea came from an unexpected place. In those days, the exposition center in Moscow was active with private international exhibitions. Most of these exhibitions opened their doors to the public on the last day for a fee. My friends suggested that we visit one to see if we could grab any free swag to try to sell at school. Three hours, and three bus, train, and subway connections later, and we were there. The harvest was good. We collected free pens, notebooks, bags, and other souvenirs branded with vendor logos. We knew nothing about these companies, but the inventory was sellable. You couldn't buy these products in Soviet stores, so the novelty alone made them cool.

During another exhibition visit, I met some kids who told me about their own business—trading Soviet souvenirs to tourists for solid currency. Foreign currency sales was another league, but souvenir sales sounded like a realistic business idea to execute. We lived on a Soviet military base, so we had access to the military uniform store. Usually they only served customers with a military ID, but no one ever asked the kids who lived on base. We could get the goods. We just needed to find a way to sell them.

That summer, at age eleven, our little gang of military kids started a commercial enterprise so that we could afford the luxuries of life that our parents could not. We lived three hours away from Moscow's Red Square where the rare American, French, German, and British tourists could be found. We pooled our childhood savings together and bought the famous Soviet fur hats with the red star, military belts with a Soviet star on the buckle, and other seemingly exotic products that might interest these tourists. We chose only easily carried goods because our sales model presumed buying and selling on the go.

Everyone in our group had clear roles and responsibilities. Three sellers would communicate with foreigners using a few English and German phrases as well as body language. Two school boxing champions would protect the sellers from other gangs selling nearby. Our competition wasn't happy about us moving into their territory. We had to find the right balance so that everyone still made money and no real fights broke out.

The sales process required a certain talent, since we couldn't display our goods at a booth. Our enterprise looked more like a secret service operation than a business. We either carried our goods in bags or wore them, approached tourists, and monitored the surroundings to make sure that we weren't noticed by the police or KGB agents, the Soviet secret service. Talking to foreigners was against Soviet rules. KGB agents wanted to prevent Soviet citizens from talking to foreigners. Some of those conversations turned into espionage allegations and real jail sentences for thousands of people. Despite our commitment to make money with our enterprise, we all agreed on safety first— avoid problems with the police and the KGB and avoid direct confrontation with other gangs.

Our first round brought no sales. We focused on learning the logistics and making ourselves comfortable talking with foreigners while watching out for agents. Our second attempt made the first sales record. We knew people were carrying one, five, and ten banknotes, so we quoted accordingly. An American guy bought a gray fur military hat with the Soviet red star for ten bucks. That was a lot of money to us. Because foreign currency was so hard to buy, there was a massive black market to trade for Soviet rubles. It's the same in many African countries today—the black-market currency exchange rate is ten times the official one. We exchanged those ten US dollars for one hundred Soviet rubles, which, minus a 20 percent commission to the exchange agent, left us with eighty rubles—close to the Soviet average monthly salary! Not bad for eleven-year-olds in one day of work. We built a routine to do this once a week and reached the point where each of the five members earned up to one hundred Soviet rubles per month. That was enough money to fulfill our needs at that age, such as buying expensive sneakers and inviting a girl to the cinema and ice cream. It was our first experience of financial independence and life standard upgrades.

What's so special about a group of kids starting a business? In the Soviet Union, entrepreneurship in any form was illegal. If the police caught you running your own business (let alone with illegal currency), you'd receive a criminal prison sentence of up to fifteen years. Many successful Russian businessmen in the early 1990s had spent the previous decade in jail because they tried to run private businesses. Everyone in the Soviet Union was obliged to have a state-sponsored job. Like self-employment, unemployment was also illegal. However, these rules applied only to adults and children fourteen and up. Technically, they couldn't send eleven-year-olds to jail.

We thought we were safe—from the Soviets, that is. Not from my father, who was pursuing a military career. After an especially busy day in Moscow, I didn't get a chance to exchange my Deutschmarks and British pounds. He found them in my room that night and drilled the truth out of me. He was less than thrilled with the way my friends and I were saving up. To my father, it looked like his son was on a trajectory to become a criminal and ruin his life. Moreover, his career was at risk of termination if I got caught. He told me to quit the business and stop jeopardizing the family.

With that, I had to end my little enterprise, but it didn't kill the feeling that I deserved better. It made me realize that I wanted a different life than Soviet norms allowed. Under their rules, I would never own my own business or have the life standards I wanted. I wouldn't be able to travel the world—basic human rights violations. Of course, I didn't know what that meant when I was eleven. I simply felt that it was wrong to be restricted from doing what people in other countries could do. This feeling became real at the age of fourteen when the middle school graduation ball came. I wasn't going to go, as I simply had no proper attire. My family wasn't able to buy me a suit. But I was lucky to borrow spare black trousers and a white shirt from a friend of mine, whose family was wealthier than we were. Having fun with friends before my departure to military school made me commit to never being poor. The fear of poverty is still one of the key drivers of my success. It has an official name, peniaphobia.

By 1990, the Soviet Union was in free fall. The predictable road map of the Soviet citizen was gone. Russia legalized small business entrepreneurship. The first real business opportunities opened up. Unfortunately, liberal thinking and entrepreneurship weren't part of my culture growing up. My parents focused on saving, not earning. They always told me to get the job, keep the job, follow the rules, care about public opinion, and choose the least risky option, whatever the opportunity was. That wasn't their fault. Generation after generation of Soviet living meant they never knew what freedom to choose your destiny felt like. They had no business acumen, so they had no chance to be a role model for me. As the Iron Curtain fell and the Soviet Union collapsed, my parents had to restart their careers from scratch in their forties. Old skills were useless in the new world.

Having little life and career guidance from my parents, I was influenced by friends as a teenager. We were the ordinary hooligans that every public school around the world suffers from. We bullied other students, provoked teachers, and planned to drop out of school, get low-wage jobs, and tell our parents to back off. My friends and I were poised to go from students to criminals.

Everywhere we moved, my peers kept me on a dangerous road. Most guys around me abused alcohol and drugs. Their habits frustrated me. I knew this wasn't the life I wanted, but I had no idea what I *did* want—or how to make it happen.

Most of my friends would be dead by twenty-five. Half were killed in street fights or targeted murders. The other half went to jail or died of overdoses. That would have been my fate if I'd stayed in that circle.

I felt like I deserved better. Yet it was impossible to have other friends growing up. After a few years of getting into trouble, my parents decided that military prep school would be the best alternative for me. Being around the wrong people would only drag me further down, and military school was the only other place I could go. I left the school of hooligans and prepared for my future military career instead of high school and university. Joining the army was practical. My family wasn't able to afford to pay for college, so getting a brilliant education for free made sense.

My high scores at military prep school got me into the Military Space Engineering Academy in Saint Petersburg, Russia. After five years of study, I'd have a professional education. Then I'd have a job in the military service for five years after that. But after my fourth year at the academy, I knew a military career wasn't for me. Officers' families lived in poverty worse than that of the late Soviet Union. Captains and majors had to take side jobs for which they were overqualified just to pay the bills. Study for years to live in poverty? That was not what I wanted.

I struggled to get any guidance in my circle in the military. I finished my five years of study and graduated in July 1998 with no plan. The global crunch hit Russia in August, and Russian currency dropped in value four times over. The economy collapsed. For the first time, I faced real competition. I was a fresh engineering graduate in a job market full of experienced candidates ready to work for a fraction of their previous earnings. The most famous joke of 1998 was "Having a cell number isn't cool. Having an office number is cool."

Every day for six months, I left the small town I lived in, rode two and a half hours to Moscow with two connections between bus, train, and subway, and knocked on doors. A friend of a new friend referred me to an IT distribution start-up. I wanted a job so badly that I convinced the GM that I had experience in international logistics. Within two weeks, it was obvious that I had no clue about import operations and freight booking efficiency. Instead of firing me, they asked me to help their purchasing department with supplier profiling. During my second week in this new position, I got them a distribution deal with Yamaha. The manager's response? "Guess you're not as dumb as we thought. I suggest you be a purchasing manager covering the lines no one has time for and see if it works."

It did. I was lucky. I had good role models there. Everyone was just about three to five years older than me, but they acted like self-made masters of the universe—finding awesome deals, creating new lines of business, looking cool, acting cool, and earning a lot. This worked for me for a while.

After six months of this, I started to feel like I deserved better again. I wanted to work abroad to gain international experience. During my year and a half at my first job, I'd developed the skills to white label almost any technology product. We'd built two private labels, and I'd operated both of them well. Word of mouth led to a call from the CEO of ASBIS Enterprises, a pretty big international IT distribution company in those days. He offered me a job on their international team in Prague to build a new line of business from scratch. This was an opportunity to experience what I was hungry for—travel the world, live like a local, speak other languages, and, most importantly, meet and learn from great businesspeople every day.

The familiar I-deserve-better itch made me quit my successful career at ASBIS two years later to start my own IT trading company. We were profitable from month one and accumulated great earnings, but we failed to scale our trading project into a full-fledged business. I had to shut it down and accepted the job offer from Seagate Technology in Russia to begin my journey to CEO. I am still staying humble and running naked truth assessments on myself every time I feel that I deserve better. This is how my every next career move starts.

Every new chapter of your career requires a new you. You need to decide if you are really up to this new challenge and if you have the required abilities and hunger—or if you just expect success to happen because you did something great in the past. It doesn't matter how long you've been on your career journey; the world still doesn't owe you anything. Deal with it and help your team to deal with it as well. Every next career step requires a new mind-set and a new circle. But finding your new circle isn't a simple exercise with obvious instructions. How do you upgrade your daily interactions and become the person who deserves the success you desire? You don't have to search for people who inspire you. If you keep your eyes open and listen, you'll meet them when you need them.

Tip 1: Identify Your Values

Everybody lives according to their values. Our values determine what we get (or don't get) out of life. If you're dissatisfied with your career, your business, or your relationships, look in the mirror. Every so-called dark stripe can be broken down to the chain of events and wrong decisions or actions taken. What do you really value in your life? What principles make your decisions truly yours? Do you want to be remembered as a nice person who never rocked the boat? Or as a disruptive high achiever who made a positive impact? Figure out what it is that you value, not what your peer group values. You've probably adopted their values without even thinking about it. Living other people's values will lead you straight to mediocrity. My friends back at school valued rocket-fast success with alcohol, drugs, and troublemaking as side effects. So I did, too—until I saw where that led.

During my early years in international sales, my boss put mediocrity in black and white to his team. "Do you want to make a difference, or are you happy to be average?" he'd ask at every business review. We all wanted to make a difference. Being the one who scores feels awesome. Seeing others score is inspiring and raises the bar. When someone gave excuses for not hitting their sales goal, our boss cut them off. "Shut the fuck up and sell this shit," he used to say. And it worked for us. It's hard to imagine addressing low performance this way today. Everything and everyone has gone soft. But it worked well for us. It kept it clear that we were there to win.

Shared values are a must-have framework for any business. It's the most secure way to choose a business partner, accept a job offer (or not), and hire direct reports whose performance you depend on. When you think about shared values, be specific. Achieving a sales target is not necessarily a shared value—it's a business objective. Making lots of money is not a shared value—it's a life goal. Making lots of money by running the extra mile for your partner, knowing that they do the same for you— *that's* a shared value. Building consistent business processes and robust sales models—these are shared values. Committing to transparency, integrity, and work ethic is a shared value. Building a legacy is a shared value. Building a business for the purpose of making money, enjoying the journey, and being proud of what you do together is a shared value.

It's sad to see so many businesses confuse their values, mission, and marketing objectives. If you say that your company values are saving the world or building a better future, you're wrong. Those are marketing statements. Values make a complex business consistently efficient today and resistant to the challenges of tomorrow. Values are always down to earth— like work ethic, integrity, or a clear intolerance of dirty politics. Be honest and specific about what you value in life and business. Get vital decisions aligned to them. When it comes to employment, verify that your personal values are aligned with the employer you want to join. This builds a foundation for productive cooperation in the future.

Values impact the hiring and partnering process more than people think. Karma says that we get what we deserve. I suggest replacing this simplistic idea with Frolov's law—we get what our *values* deserve. If you value integrity, you will live an upstanding life that attracts positive opportunities. If you prefer to waste money and time on vices like alcohol, then you, like many people I know, will care about nothing when the bottle runs dry.

Tip 2: Improve Your Circle

The kids I knew who ended up in prison probably deserved it. If I'd stayed around them, I would've done things that merited jail time, too. If I'd kept hanging out with my laid-back, heavy-drinking buddies after the military, I probably wouldn't have had a career. I wasn't *entitled* to a good job just because I fulfilled my duty to my country. Why should anyone give me a job? I had no experience. When I finally found a start-up willing to give me a chance, I felt *grateful.* That gratitude motivated me to earn and prove myself worthy.

I've made an interesting observation. The absolute majority of the top-level managers I knew ten years ago are out of my radar today. I know they're all in good health and still do what they used to do. The reason I don't see them anymore is that they stayed where they were, and our orbits no longer cross. I moved forward. It's clear to me that the people who inspired me to do more at every step of my career made this difference.

There is a phenomenon of how the wrong circle becomes your cage. Anyone who lives in the corporate world used to experience it. The corporate environment incentivizes people to make friends at work, and the fastest way to do it is usually to criticize or hate others, to complain or share disappointments about things going on in the company. It is easier than standing up and challenging what you dislike or addressing your efforts toward fixing what makes anyone unhappy. These groups are having these conversations during coffee breaks, lunches, or after-hours drinks. People are influenced by it much more than they might think. It's corrupting the way people see opportunities, the way they address problems, and the way they start to see their business in perspective. Have you heard that misery loves company? This is the simplest way to describe it. What happens when people create a circle that mostly sees things in dark colors and criticizes more than they do? They stop growing. There is no benchmark or role model for them within this nine-to-five circle, no time frame for success to develop. It is hard to realize that you are surrounded by negativity because it becomes a part of your behavior. I have found myself in this situation a few times in my career. I've realized that it is all about how fast you can escape this circle without harming anyone.

Spot these wrong circle red flags:

- Are your friends and colleagues complaining about how everything is wrong and unfair, or do they escalate problems constructively?

- Are they executing with excellence or pointing fingers at those who they think stop them from doing it?

- Are they making a difference for the department and the company or barely meeting expectations?

- Can you list what inspires you about them and what you can learn from them?

Here is what always worked for me to change my circle:

- Get a mentor inside the company who can elevate your vision.

- Identify the competencies you are missing and ask to work on a project that will help you develop them.

- Spend more time with successful people outside your company with a purpose and agenda. Share your struggles with daily operations and absorb the feedback.

I credit my changed mind-set to the first time I tried networking with a purpose. I upgraded my circle, and in turn I upgraded my outcomes. That doesn't mean you should abandon all your friends. If you have friends who enjoy football or fishing or nearby activities, that's fine. Yet you always need a circle of people who mentor and inspire you to think differently. You need to share values, not just hobbies. Do the people you spend time with have dreams as big as yours? Do they pursue ethical ways to earn money rather than questionable ones? When you share values with a circle of good, smart people, you become a good, smart person, too.

Since then, whenever I've wanted to move up in my career, I've changed my circle. I chose to hang out and network with management, top performers, and business owners. These people helped me grow. How? Instead of running my mouth complaining about deserving better or bragging about my minimal accomplishments, I listened. I found people I wanted to be like, and I took their advice. Simply overhearing their conversations opened my eyes. They had ambition and work ethic to match. Their influence inspired me to go the extra mile and to live a life of integrity.

At my first job, I saw real professionalism for the first time in my life. Seeing them make a difference with their work and stick to high standards of how the job should be done, I adopted best practices fast and altered my vision of what I wanted in the short term. They formed my new circle. They pushed me forward. The values we shared were simple yet brutal: *Outwork anyone to make good money. If you screw up, go fix it. The job must be done right. Mediocrity sucks.* Interestingly enough, those values have remained in my circle for over twenty years already and are still going strong.

The productive relationships we built around our shared business objectives have grown into solid friendships based on our shared values: *Dream big. Work hard to make it. Put work ethic on a pedestal.* The right circle is like a collective brain that makes you all smarter. When I've felt stuck in the corporate world, I've changed my circle. I've met down-to-earth businesspeople inventing new products, creating new business niches, and living in firefighting mode every hour of the day. Several of my business ventures came from these relationships, including the Urban Innovation Group start-up.

How many times have you heard complaints from people who hate their jobs? The bigger the company, the bigger the complaints. Many of these people explain away their poor results or absence of purpose and blame colleagues, bosses, and company strategy. I see a tremendous group of experienced, professional, high-potential people who can make a difference. But they limit their potential by complaining about their bosses. Corporate politics is a poison that makes companies waste millions every year. It creates a toxic environment that makes good professionals give up and become mediocre.

What surprises me is that the professionals who suffer from a toxic environment often stay at the job they hate. You are not a tree! If you're unhappy with but unable to change the circumstances, just move. The market is always short on solid performers. You can always find a better job if you are good at what you do. If you embrace this freedom, you don't need to sacrifice the truth or your self-respect to keep working for people with ethics issues. When you know you can find good work elsewhere, you can be objective in your decisions, more efficient, treat people well, and never mess with politics.

Even if you're good at what you do, you still need financial security to truly have this freedom. What if you decide to move with no job offer on hand—or you're fired? Plan for the worst-case scenario of how long it will take to find a job you want. The higher the position and the more responsibility you have, the tougher the job search. I always recommend that you have a cash buffer, or better yet, a mix of cash and passive income to cover twelve months of expenses and to keep up your lifestyle.

Should you maintain your lifestyle, though? Is it realistic? I'm not suggesting that you never go to the movies and eat only ramen. But so many people live paycheck to paycheck because they've decided to buy a house and car they can't afford, they took a high-interest education loan, or they're buying designer clothing and bling. I'm not criticizing you if you're buying the finer things of life. As a guy who couldn't afford what he wanted in childhood, I grew up desiring exotic cars, international travel, and expensive watches. Nothing is wrong with this either, as long as you can afford them with no impact on your investments and savings buffer. If you feel miserable at work and can't find the job you want, you'd better accumulate savings, not buy a Range Rover.

Your circle is the best accelerator for your career. If you hang out with top earners, learn a little, but pick up an excessive lifestyle, you misunderstood the concept. Your circle should inspire better achievement through hard work, practical advice, and role model behavior. Your circle can become your MBA class. Your role models should share stories of failure and how to turn them into success. If you don't experience this in your circle, rush to change it.

Tip 3: Sacrifice More to Deserve More

The fastest way to determine whether you are entitled is to ask yourself what exactly you are sacrificing to get what you believe you deserve. If you believe that you are too good to work for someone and that you should quit your job and try new things yourself but are refusing tough or uninteresting jobs in a situation of low cash flow, you have an entitlement problem. If you work for the same company for ten years and expect that raises or promotions should come automatically, you have an entitlement problem. If you have accepted a position that is not your dream job and don't work your face off because of it, you have an entitlement problem.

No employer owes you anything beyond the salary you receive. No customer owes you anything just because you are an expert. No business partner owes you anything just because you believe you are more experienced, clever, or hardworking.

Have you heard the expression that everyone wants to go to heaven, but no one is ready to die for it? That is exactly what I mean by an entitlement problem. What are you willing to sacrifice to earn what you want? Would you sacrifice your free time to learn something new or try something different? Do you really expect something nice to happen to you just because you deserve better? Have too many motivational memes told you that if you wish badly enough, you'll get what you deserve? Can you invest your free time into networking with others who share your values so that more like-minded people know who you are? Or learn something practical that will get you closer to what you deserve? Would you sacrifice aspects of your lifestyle to invest in your side hustle? Are you ready to sacrifice half of your current income to start your own business and work hard to get back to the same earnings level five years from now? Do you really need a watch collection or luxury car at this phase of your life? A house that stretches your budget? Jet set-style vacations and weekends? I used to buy rare watches that depreciate in value immediately because I thought I deserved them. Not a smart investment. I'd have been better off investing more in the stock market or real estate and building my watch collection later.

Will you sacrifice your own money to go after what you value, even if there's a chance it will end badly? I pulled from my savings several times to start new businesses, which failed. Even today, I give up investment opportunities to cash in on new, risky business opportunities with great potential. I become an active partner in new ventures. Sometimes they pay off. Sometimes they don't. It's all part of the game. Yet I'm never sorry I took a chance. That's one of the things I value. Taking a chance and going after something exciting that just might make a difference.

If you've found yourself feeling like you deserve better, I hope this chapter was your wake-up call. If you want better, *earn* better. Work hard, choose your peers wisely, and sacrifice your comfort. The world doesn't owe you a thing until you do and deliver.

The absence of proper business acumen misleads many good, hardworking nine-to-fivers who are inspired by persuasive motivational content and dream to become their own bosses or start to earn more. If you're going to jump into the dark, cold waters of entrepreneurship, risk management and a strategic work ethic are your two life preservers. Otherwise you will freeze and sink like Leo in *Titanic*. Risk management and hard work are rarer than you might think, probably because of a mismatch between expectations and reality. If all you hear are success stories, you come to believe that self-employment isn't that risky or tough to accomplish. Then you underestimate the skill set you need going into your venture, and you end up working on the wrong thing at the wrong time. Too many people confuse positive thinking with their naivete.

It's never too late to start working for what you believe you deserve, but that's assuming you already have what you need to begin. If you look at someone whose company earns millions in profit after starting from scratch, what you won't see is the market knowledge, business acumen, long line of failures they've learned from, and time spent to ramp it up to a sufficient cash flow level to pay the bills. Remember, the people you see making it today started with excellence in their craft, a clear understanding of the business processes and finances, and the capability to attract customers. Most of them learned and mastered these in a nine-to-five job.

Tip 4: Be Worth More to Earn More

Most people believe an illusion about how much they're worth outside their current jobs. It's all relative for the folks just starting their careers, but after ten, twenty, or thirty years, it's different. On the one hand, the easiest way to increase your skills is to change companies, but this is a merit increase, not necessarily an increase in income. When you have a solid career in the same company for several years, you start to see the world through your company's universe. You feel like you're big stuff in your current job, but that doesn't mean you have big value outside that company.

I recently grabbed a beer with a guy I used to work with. He said to me, "I think I'm undervalued. If I decide to move to the US from Europe, a guy like me is worth two and a half million a year."

"You're fundamentally wrong," I told him. "If you reverse engineer what you do now for another company, your salary there would be half of what it is now. You have more than ten years at your current company and a success story driven by organic market growth. You have built a reputation at the board of directors level. This is why you're making such good money in your current role. Most of your knowledge is about how to be successful in your current company with a certain set of products and sales models and how to navigate the internal politics and influence the owners. That's not scalable outside that company at the level you're looking for. Most employers are looking for C-level executives who can either handle crisis management or drive transformation of the business in tough environments. You're letting your achievements at one job multiply your perception of your value. In your current role, you have an incredible aversion to missed shots because of your credibility internally. This won't be allowed in another company."

Too many executives confuse talking like an entrepreneur with acting like one. A real entrepreneur is called an entrepreneur; all others are wantrepreneurs. Leaders with the entrepreneur mindset have the capabilities to run a corporate business like their own in terms of verifying risks and opportunities, investing responsibly, and executing the road map with commitment. Within three months of announcing a new initiative, line of business, market segment, or product, you can see progress on the road map. In three months, wantrepreneurs are still proposing, discussing, promising, committing, and requesting but showing no progress.

The revenue and profit you directly create is a primary metric to calculate your worth. Hands-on management becomes a vital skill to preserve and increase your value. Many strong performers fail when changing jobs because they used to have several people who did the job before they did. Sales is a good example here. If you're hired at an established business where the sales channel and customer base are up and running, and you deliver consistent results, it doesn't mean you'll be able to solve problems at companies with performance trouble or build a start-up's sales from scratch. There's nothing wrong with having no crisis management or start-up experience. To deserve better than what you have now, you need to obtain these skills. If you manage complicated, multidimensional sales models and organizations as a field coach who can hire right, drive the onboarding process, and lead by example, you have a higher value than your peers and managers—especially those who are more Excel warriors than active business leaders.

No two career paths look the same, and that is quite all right. My dad started his career over at the age of forty. He had a very successful career in the Soviet military. But when the Soviet Union collapsed, he had no choice but to resign. Because my dad's background was space communications engineering, he was extremely good with numbers. Before the Soviet Union collapsed, real estate was entirely controlled by the government. Even though my dad had no context for the real estate business model, he got a job in the free world as an analytics guy. He basically started his life over with no business acumen and zero understanding of the capitalist world. He never had a mind to start his own business, sell something, or make money outside a job, but he was self-aware of his analytical mind.

He got a job at a well-known real estate company and worked his way into middle management. Even today, being retired, he has a part-time job because he's an extremely active, well-developed person. He cannot just work in a garden; he's out consulting with huge corporations on their future real estate ventures, business models, property regulations, and so on. Hard work helped him make a consistent career for over twenty-five years in real estate development.

Let's face reality. Everyone has to be a salesperson today. Whatever you do—sales, marketing, finance, or product development—sales is a key driver of any process in business today. This skill set will let you become a valuable asset for any business, especially one looking for a leader who can do more with less. Leaders who can identify obstacles and apply a quick fix as well as a long-term strategy are valued the most.

The complexity of the business model at your current position determines your value in the job market. A high-revenue business with a simple business model will always discount your value compared to complex business model professionals. If you do $100 million a year selling a product in a box through two distribution partners, you're the champion of managing two distributors. If you own relationships with their key customers, your value increases. If you make the same $100 million but sell complex products that require presales objection handling and consulting, both sales channel management and direct sales capabilities, or deep knowledge of digital sales, marketing, and ecommerce operations, your value just skyrocketed. If you run $100 million across five lines of business, you will always be more valuable than your peer who runs the same $100 million in one line of business.

Another area that impacts your worth is leadership. Managing people is the toughest part of any managerial role. Hiring the right people, managing low performance, and creating a framework that makes employees deliver stunning results is the value a lot of managers are missing. Everyone is ready to get promoted, get higher status, get a bigger package, and get a nicer office, but not everyone is ready to work for their employees to make them successful. Working for your employees doesn't mean you do instead of them. It means you create the framework for them to be successful, not just make you successful with the results they are expected to deliver. You act as a filter to prevent your team from being distracted. It takes courage to share wins with the team and to accept their missed shots as your responsibility. People management capabilities are valued in the number of people under your direct leadership as well as the number of indirect reports you influence. The more diverse group of people you manage with different roles, the higher your value.

The number of business processes that you manage also impacts your market worth. There is a different value between managing two processes across sales or twelve processes. Your capacity for bigger tasks as well as career potential is higher the more you've proven you can tackle.

Another way to calculate your worth is your geographic coverage. Previous multiregional or multinational roles are more valuable than local and domestic ones. If you manage a $10 million a year business in France, your value in Southern California is discounted heavily. You're missing local specific knowledge and market dynamics understanding.

In summary, if you want to increase your pay, increase your value. You now have practical ways to do that. Remember, in every role you get paid for the value you add, not for the time you spend. Don't take for granted that this will happen without effort. If you feel that you deserve better, *earn* it. What tasks must be accomplished when you have your dream job? What value do you need to add to get paid what you want? Figure out what skills and experience you need and gain them now. Some people assume that they already have what it takes to climb the ladder. They pursue a pipe dream.

Either that or they follow bad advice about "niching down." If you're going to be a niche-down person, you might have quick gains but will likely fail in the long run. Just think for a second. How can you lead a large-scale courageous transformation if you're familiar with only a specific group? Even if you're a well-paid specialist, you limit yourself from delivering big results. This applies not only to large corporations but also to small businesses. When I led Urban Innovation Group, we had to reinvent our business model every quarter. Our product was early to market, so we had no choice but to find other ways to monetize before the time to market came. In my marketing automation start-ups, we used to adjust our go-to-market strategy once a quarter as well, sometimes even more often. You can't do that if you know only one niche in your company.

We all want to live a happy life, contribute to the economy, and pay our bills. For a lot of people, it's better to do that through a stable job, and that's perfectly fine. There is one thing that many people have missed recently—any job that pays the bills is a great job. Jobs are not separated between the good and the bad. The real line is between well-paying jobs and poor-paying jobs. Between your excellence at the job, the difference you made there, and your failures at the job, poor deliverables, missed targets, and lack of recovery, is it better to stay or to start your own thing? I do not know. But when *you* know who you are, you chart the best, most productive way to spend your time.

Tip 5: Quit When It's Time to Quit

People who feel entitled to the outcome they want rarely understand sunk costs. If they invest effort, money, and time into something but do not see results, what do they do? Keep investing until it does one day. Maybe. This is a formula for ruin. If something is not working out, it's not working out. There is nothing wrong with changing your mind. For example, it's OK to run a new business for a year, realize it's not going to fly for solid reasons, and shut it down. Every business has an entrance cost and an exit cost. These are a part of the journey.

If people aren't buying what you're selling, it's time to reconsider if you're doing the right thing. It might not be the right time to market, or the product you sell is not good enough, or you are not good enough to run this business on your own. Not everything is sellable at the price point you expect. It's all right to change your mind, change the strategy, or shut it down to start another venture or to go back to full-time employment. I've said it before—there is no such thing as a wrong career move anymore. Personally, I hire people with as many diverse experiences as possible. From my perspective, people with self-employment experience are a company's most valuable asset because they are more responsible and agile and better understand how to drive business lean.

If you don't see the results you desire over a long enough period, if you receive solid proof that the business isn't turning out as you expected, or if you realize that you won't be able to sell at the level your P&L requires, you're in the right position to quit. At the end of the day, the market decides. Does that mean you don't have what it takes to be an entrepreneur? Not necessarily. This may surprise you. Entrepreneurship is not a profession. It's a mind-set. A lifestyle. Many corporations today nurture entrepreneurship as part of their culture. They offer opportunities to freely bring their ideas to life, take risks as entrepreneurs inside the company, and make a difference for everyone. Pursuing entrepreneurship on behalf of an employer is a great middle ground.

Quite a decent job has been done over social media to convince people (especially millennials) that the nine-to-five is wrong. Traditional employment is not wrong. Having no money is wrong—poverty cannot buy anything. It's insane how many people I know who say, "I'm too good for this shit," quit their well-paying jobs with a poor business plan, and go broke. Entitlement leads them to believe in the illusions of guaranteed success, start unprepared for the real world, and ruin themselves.

Be realistic. Respect the position you have now, learn from it, and always mind the time of your career journey. That's how you prove yourself worthy of the success you desire and eventually move up.

CHAPTER

Fake It Till You Make It:
Terrible Advice about Confidence
(And What to Do Instead)

"There's nothing wrong with faking it till you make it," we're told. "If you're a good person with a bright idea, a beautiful product, and a commitment to make it happen, there's nothing wrong with projecting success to attract success."

Remember the last young entrepreneur you saw showing off on social media? Maybe it was a picture of their Lamborghini. Then a Rolex watch. Then an expensive champagne party in the club. That's consumerism, not success. It's fake. These people use social media to show off their massive purchases to hundreds of thousands of followers. In their jet-set lifestyle, they buy new supercars and travel the world. I also know that, to afford these luxuries, they often spend money they don't have. Banks won't finance some of these people anymore because of their debt. I personally know several of them. Others spend all their profit to look bigger, better, and more successful than they are. They show off for the sake of showing off, doing their best to impress others, expecting something remarkable to happen someday. Faking it, never making it.

Social media does a good job of beating up these bullshitters. Most followers call out the people pretending to be better than they are. This is basically anyone saying, "I will be a millionaire before I turn thirty!" Most of these entrepreneurs and self-made CEOs of one-man companies aren't in a serious business anyway. They don't really interact with the economy. What they mostly produce is content. This is totally fine. People are making money from advertising proportionate to the number of followers they have.

What has a bigger impact (and not a good one) is when "fake it till you make it" lands in the corporate world. Everybody does this, and nobody warns you not to. In fact, we're told we're *supposed* to.

Faking It, Never Making It: How Overconfident People Ruin Businesses (And What to Do about It)

Every employee in the world knows how "fake it till you make it" sounds in corporate. It means "Promise big. Show quick wins." Quick wins cut corners today and cost more later. In the majority of cases, once you're hired, you're expected to show an immediate impact and to promise a bright future. If the situation is desperate, you're expected to be Harry Potter, wave a magic wand, and fix everything. Most companies who ask me for referrals have these expectations from candidates. That's not how reality works.

Successful companies work like a machine. Machines can't fake a function. Either they work or they don't. There's no pretending. A quick fix might make a problem seem solved, but when that fix breaks down, everyone is worse off. Yet I've seen people *pretend* to be top performers even with poor deliverables. This behavior pattern exists in almost every company. They overcommit company resources. They overforecast sales. They set huge expectations and live on their empty promises. All to look good to their superiors. They hope sales will increase because of luck or an organic demand miracle in the last quarter of the year. Every other Christmas throughout their careers they've seen a revenue bump, so why would this year be any different? They count on the past to predict the future, jeopardizing the entire business.

It's not just employees and middle management who are overconfident with no actual basis. I've also seen company leaders create a fake reality. They tell everyone how special their team is and how blessed they are with their mission. They promise enormous results to impress shareholders and latch on to industry trends to seem relevant. They don't get their hands dirty. They don't transform the business model when fast-changing market conditions require it or start new lines of business to get ahead of demand. They paint a bright future and an aggressive growth picture with no basis, all to look awesome at important meetings. Sound familiar? Capable leaders work for results, not praise. Usually, people who fake it would rather get credit for promising big than working hard to deliver big. All hype, no help. In the end it generates nothing but the bitter taste of disappointment and poor financial results. Fake performance is not a sustainable business model.

Technology drives the evolution of everything we interact with. It makes everything more convenient and productive. However, disruptive technologies are cutting more jobs than they are creating. New jobs require different skills than did the jobs that are becoming obsolete. Many more professions that require you to work hard to pay the bills today will also disappear in the short- to midterm. Many business models that allow companies to hit the P&L objectives today are melting as I write. Our fast-changing reality makes a lot of leaders lose relevance. And they start to panic. Many have to imitate progress. They do what worked years ago. They want to calm down upper management because they can't explain the root cause of the problems and they can't execute a fix. Sometimes managers can't evolve with the changing times and attempt to replicate past successes that aren't working. It's like doing the same thing over and over again and expecting a different result.

This type of manager is soon to be obsolete. Transparency based on data integrity links the global ecosystem. It's harsh, but that's reality. Do all managers comprehend this? Not really. Many prefer to fake a smart plan and pretend they're able to drive change. They can't. This fakery costs the global economy billions of dollars every year. Even as sales fall, C-levels have to listen to these managers, who advise spending profits on pointless tactics while having no strategy. This is where past wins drive today's promises and mislead so many decision makers. Meanwhile, those managers expect issues to sort themselves out. Others have no choice, as they might have gaps. They run the business and expect results to materialize from nowhere. This is not just naive, it's criminal.

Why do so many previously successful managers do it? Why do they fake competence? It's better than doing nothing, those managers have told me. Why don't they analyze the root cause and fix the problem? In many cases, they don't know what to say. Starting and fixing things is hard. Imitating success is easier than earning success, especially if past wins give you credibility. Humans are weak. We always take the easiest path—it's natural. It's easy to avoid seeing the big picture and its associated challenges because it is a greater responsibility. Why take responsibility when you can protect your job for another year and figure out what to do later? Because everybody lives month to month, quarter to quarter, fiscal year to fiscal year. There's a certain career path that gets you rewarded and promoted. To keep on that path, people fake it till they make it—or get fired. This is how most companies work.

When I managed worldwide sales for Kaspersky, I had a classic fake-it-till-you-make-it example to deal with. She was one of my employees, a direct report who looked over the largest sales territory. She had a high position and excellent communication skills. She could convince anyone who wanted to hear a success story that success was there, especially when Kaspersky became a toxic company for many customers and partners worldwide due to the issues the company faced in the United States and soon in the rest of the Western world. Sales numbers slipped and core partners terminated as bad publicity soared. Every week brought more bad news, from an article turning up the heat to another country banning Kaspersky products.

Given this situation, many senior executives wanted good news. People always look for something positive in a bad situation to help us through the tough times. "We're not giving up under the geopolitical pressure!," "We didn't do anything wrong!," and similar statements became the norm inside Kaspersky. It was easy to fall into endless motivational activities with no actual plan to remedy the situation. Looking after aggressive sales and margin targets, I had no luxury to let sales teams waste time. Top-line revenue numbers were aligned to payroll, offices, and other fixed costs. We had to deliver the P&L numbers no matter what. It was clear that talking about how bad the situation was and spending time to motivate ourselves wouldn't take us far. Execution excellence is the answer to any external challenge because it's the only thing we can control. I needed my team to share my vision and execute as required despite the company's desperate situation.

That brings us back to that one particular salesperson. She set high expectations when she joined the company. Her job was to fix sales performance problems in spite of massive external pressure. That's the very reason the last managing director was fired and she was hired. Sales, market share, and business productivity metrics were failing, and the ex-leader didn't know how to fix it. The company's product portfolio was struggling to match the rising competition, which didn't help. In the face of external factors, many companies see longtime efficient business models crumble. Kaspersky was no exception. The organizational structure and sales model didn't fit the new market reality and the way customers were buying. The company's value proposition and sales channels weren't adequate for the fast-evolving market. Perhaps declining demand meant that brand value was diminished. My newly hired direct report claimed that she understood what was wrong and could crisis manage Kaspersky's business into a solution that worked.

Unfortunately, that never happened—and for a simple reason. We needed a well-executed get-us-out-of-the-crisis plan, and she did not bring one. Have you heard the saying, "You can't make an omelet without breaking a few eggs"? Anything that pisses people off is tough to execute. Crisis management, if you do it right, makes many people unhappy. It has to—you're pushing people out of their comfort zones. If you're going to fix the problem, get sales and profitability back on track, and reposition the business for consistent growth, get ready to get your hands dirty.

I knew that if my new employee did her job, she'd find that up to 20 percent of her team was redundant and that she had no budget to hire more. On some teams, ten people did the job of one. If she did her job, she'd find that her partner channel didn't give her the scale her sales target required. She'd find that her sales management processes were confusing and inefficient. She'd find marketing campaigns burning money with questionable returns. She'd find enormous opportunities for growth. She'd warn executives about a sales model generating questionable outcomes. She'd execute proper crisis management. The entire company's performance depended on getting the largest sales territory on track. And she'd build a simple recovery plan and focus on execution because we still had the time and resources to act efficiently and ethically.

If she did her job. She did not. Instead, she decided to replace true crisis management and hard work with a constant projection of success. Although the results weren't there, she invested most of her time and effort into corporate communications, such as speaking at conferences and events and to the media. She took the stage at meetings, declaring future success with passion and commitment. Most future achievements were planned for the last quarter of the year to create an illusion of progress, as "We're heading there" was said often. Her high promises got the entire company excited, the board of directors included. We all want to hear that the silver bullet solution is real. In most cases, it's misleading. Still, it's understandable. It's easy to hear what you want to hear versus hearing that the roof is on fire and that we have to apply a fix that people won't perceive as nice. On many occasions, leaders are not looking at the details of what is going on because reality might be ugly. My employee received massive credit up front while revenue projections were in the red zone. Giving credit up front helps unwire the leadership potential but might do harm if it goes with no limits or deadlines. It's interesting to see how false promises supported by unlimited credit damage organizations much deeper than it might seem. Apart from setting the P&L forecast at risk, it impacts employee productivity and morale. When you as a leader keep promising results with no actual workaround or improvements, these loud promises create entitlement among staff. Think about it—if you blame your underperformance on external factors that you can't control like geopolitics and ignore the status quo, you create nothing but entitlement. It's easier to say that you're working hard and making progress than it is to take responsibility for underperformance in desperate circumstances and to commit to fixing it.

That team's bread-and-butter doers, those who still made their numbers in tough circumstances, realized that they didn't have my direct report's support or appreciation. They understood that her fake promises wouldn't materialize because they knew how the business worked. They did learn that it's OK to be a constant underperformer who blames geopolitical pressure while proposing no remedy. The entitlement triggered a general productivity decline, as morale plays a crucial role in this equation. The easiest path is always the preference.

At some point, every business will have a manager who lives in an imaginary world. What happens when these fake promises collapse? The fake leader points fingers. Layoffs occur one to two levels below. The most talented employees start leaving the company, feeling undervalued. However, business foundations remain broken. It is not only unproductive but also dangerous for business. The one person responsible for the catastrophe stays while the employees they were responsible for are let go. This creates a fast-growing toxic environment inside the organization. It impacts business results today. Faking performance creates a lot of side effects such as playing politics and compensating for gaps in performance by pointing fingers, which are unhealthy for the performance environment and don't deliver on promises or fix the problems that won't go away by themselves. It also kills employment values, ruins the brand, and makes it hard to attract top talent. This is what was about to happen at Kaspersky. The most important sales organization within the company was on the brink of collapse. However, we found a way to mitigate those risks and to keep the sinking boat above the waterline. That required me to take personal leadership to help my direct report's team navigate out of the crisis.

First, we assessed the existing organizational setup versus the business model that would work in the current challenging environment. With no surprise, we found enough resources to reinvest into areas where we saw potential growth. In a short period, we created an organization for growth blueprint, a comprehensive organizational structure based on a three-year market outlook (i.e., what and how customers will likely buy), with the sales model and business processes aligned. Usually such an exercise takes six to nine months and is done by a global consulting firm. Having no time and budget for that but having a consulting background and experience, my small team and I did it ourselves. It was the backbone that was altered in accordance with the local specifics. The retail organization got all new business processes that improved overall business efficiency and predictability and applied a quick fix to sales numbers. The resources we released from the retail team were reinvested into the largest consumer business opportunity, ecommerce. But the biggest transformation happened in B2B, where our new blueprint improved sales coverage and productivity more than 20 percent within the same headcount and budget. All new business processes focused on sales efficiency and scalability to prevent the largest sales territory from collapse while building a base for recovery.

The team leader was fired once it became clear that her big promises would never materialize and that the company still needed to fix the same problems. Looking back, I would say that her departure should have happened earlier. Time for recovery was wasted, and some processes became more expensive to fix. It is vitally important to challenge fake-it-till-you-make-it leaders once you are clear that they impact your business. Simple and practical action based on an easy-to-execute business plan always delivers more than any public demonstration of commitment or continuous promises to make it happen. We must be realistic. In today's fast-changing business landscape, it's impossible to survive on promises and expectations longer than one fiscal year. Margin call happens sooner or later in every company. Talented management must figure out what needs to be done and must not rest until it's fixed. Perhaps there's a simple way to eliminate this bad practice and turn it into productive execution.

Tip 1: Call Out the Bullshit

In every organization, top line and bottom line go hand in hand. If you're meeting or exceeding sales targets, you can cover expenses and invest in growth. If you're not meeting your projections, you'd better cut expenses proactively and find a way to reinvest them into the areas where you see better outcomes.

I recently consulted a well-known global corporation through their latest storm. A department in a specific territory had failed eight quarters in a row. I interviewed top management as part of the identify-the-problem exercise. I asked precise questions.

"Before you got promoted, you led this department yourself," I said. "A couple of years after you left to run the territory, the department failed. Do you have any idea what happened?"

"No, actually," one guy said. "That's why you're here."

"I'll ask in a different way. If you break down the operations for this department, where do you see a weak link? Where are your people not performing? Can you assess business process quality, the sales model, and the sales channel efficiency?" I saw all these gaps through my quick analysis, but I wanted senior executives to understand this for themselves.

"The people? The people are great. Why? Because I hired them. Just a couple of years back, they were the best team worldwide. Our business processes are fine. No need to dig into them. That would slow down employee performance. There's just no answer."

Often great results happen because a company catches a wave, rides that trend, and can't explain what happened or what's next. This is totally fine. Good luck is always a part of the game. If you don't know how you achieved your results, you won't know how to remedy failure and retain the success during the next fiscal period. That was the situation at this corporation and at many others I've been involved in across the world.

This is dangerous. Why? Let's say that you made the wrong decision but that things worked out anyway. You made your target, received a bonus, got promoted, and looked cool. All good, right? *I succeeded because we have such a solid business foundation under my leadership. We're such a great team. We can sustain aggressive growth and survive any storm.* If that's what you think in a situation where the high tide made you succeed, you're misinformed. The business game is a long-term one. What happens when the storm comes? You, the responsible manager, have little room to stop the breakdown. Imagine a group of people gathered for a meeting—an ordinary situation seen in every office on earth today. Some leaders talk a lot, review a lot, and make a lot of decisions. Many of those decisions rarely turn into actions that make a difference. They're faking these decisions until they know *how* to take action. What does it look like when companies fake everything in every department?

Sales managers say that they're making progress toward the target, but they struggle to achieve results. They may have a nice spreadsheet that projects steady performance in the future, but it lacks the details on how to go about it. Excel management becomes a great substitution for reality. In many cases, nobody knows when these numbers will materialize, who will make the crucial decisions, or who has the responsibility to execute. How many people in the corporate world believe that the toughest part of the business process will be done by a mysterious somebody who is expected to understand everything without guidance? Mr. Somebody—the biggest contributor to success.

Many employees tend to back up every average or OK result with a story about how hard it was to get and how big their role in the achievement was. This is understood. Everyone wants to be part of the success story and to stay away from failure. If you're a salesperson, you get paid for on-target results. One hundred percent target attainment is your minimum goal. If you're not overachieving in a manner where the *how* is clear, there's nothing to talk about. Excuses have no value. Even if you *are* overachieving, you're still required to be honest with yourself about why. Maybe you got lucky because the market grew faster than expected and your target appeared to be too low, or your competitors misread the market and made a mistake, or your company's awesome product sold itself. Once you understand *how* you got your results—your efforts, your luck, or both—you're no longer living in a fake world. You set yourself up for consistent performance down the road. You won't accept inadequate growth rates and start to challenge them to be realistic. You start to apply strategic solutions to your business setup. You start to be proactive to spot opportunities and red flags.

Sales is not only the area where you experience fake promise problems. You'll often find such fake leaders in marketing of all kinds. In every industry I work in, marketing struggles to produce a required level of the return on investment. It is just a fact of life. The cost of new customer acquisition keeps growing and likely will keep growing further. Being efficient in this aggressive environment has never been more complicated than now. It has never been easier to scale brand exposure to your target audience either. Budgets are increasing, and the cost of errors are growing with them. A lot of marketing teams reverse engineer other companies' successful marketing campaigns. They assume that these strategies will work for them in the same way, which is understandable. However, every campaign is unique and requires a bespoke approach. Marketing investments that are reflected in sales reports are every manager's dream. However, what happens when a campaign doesn't click as expected? Instead of admitting the missed shot, many people have to find a CYA excuse to explain why they didn't produce significant results. Recently, I saw one respectful marketing director request a $60 million advertising budget to make $15 million in sales. Questioning why this investment was needed found no answer. It was never approved. If the company grows, many marketers like him take the credit. In many cases top and midlevel management are not fluent in marketing metrics and how they influence sales in the short and long term. It is easy to be misled on the actual marketing impact. This happens much too often, unfortunately. If you invest ten dollars into marketing, marketing leaders should be able to tell you exactly what that ten dollars produced.

Don't let a fake reality damage the real world. You as a leader are in charge of determining what is false and what is true in your business and correcting course. That means having tough conversations with employees to get clear answers about *how* targets will be reached. Ask dig-deep questions. Get the facts about the actual state of business. These questions will get you started.

- Are you displacing competitors, or is it a brand-new deal? Are you competing with several other companies, or is this a tailor-made deal? Is the budget there?

- What is each prospect's decision-making process? How does procurement work? Who signs purchase orders? Who approves them? How frequently does it happen? Is it a committee that reviews a hundred requests, or do they go one by one daily through some electronic system?

- What requirements need to be met to be qualified for their decision-making? What's the deadline for that decision?

- Ignore probabilities of the deal. Probability in sales is always 50 percent. Yes or no. It's that simple. Operate within the deal stages' methodology, where every step of the deal's progress is well described and aligned to the specific deal stage. Literally check boxes throughout this exercise.

- Get educated with the marketing fundamentals. Key marketing metrics, KPIs, and deliverables must be well understood. You must become fluent in understanding how every marketing investment impacts your sales results and key sales-related metrics. This will make you invest as if you have only one dollar.

- When it comes to reporting organizational performance, make sure that you don't build your performance assessment and projection based on what your reports say. Use transparent data tools and build business processes that together give you an unbiased view. This will set you in the position of a value-added adviser to anyone reporting to you, as those people might miss your perspective because they're so busy with day-to-day operations.

If you don't have these questions answered, your forecast corrected, and upsides found, you're likely in trouble. You'll end the quarter—and the year—disappointed. You'll blame the salesperson who didn't achieve the target or the marketing person who failed to create expected demand. But, ultimately, it's your fault if you let other people bullshit you. Fix performance problems and compensate for gaps. Be proactive in driving the performance; don't expect that it will materialize by the end of the fiscal year.

These expectation management examples are relevant to everyone, whether salespeople report to you, you run a marketing team, or you do something else entirely. If you let people around you fake it till they make it, you're the fool.

In 2016, Urban Innovation Group won a project in Atlantic City, New Jersey. We soon learned that the contract we won displaced our European partner, one of the industry's biggest names, who'd won the contract with Atlantic City many years prior. How could this be? We'd been great partners in Europe, and the last thing I wanted was to ignite a fire between us. What had they done to lose what should have been an almost automatic renewal to an unknown start-up? Their US account manager didn't know about the new circumstances that prevented automatic renewal. This account manager told the company that contract renewal was guaranteed. Nothing to worry about. They opted into this sales forecast and faced an unpleasant situation later because of the account manager's fake expectations and overconfidence. The company could have bid and won a renegotiated contract, but the account manager blew it.

Now more than ever, it's important to get the facts and to live in the real world. As the next global recession has arrived, we need to put the right people, processes, and plans in place to survive this storm and to use the opportunities it opens to many businesses. How can we do that if we don't know which teams need attention or where our business model needs transforming? This is what happens when most companies hit the wall during a recession. When you're at the peak of the economy's performance, it is hard to start preparing for the downturn. At the first signs of slowing demand for their products or services, businesses must adjust. People who make nice promises and give overexcited forecasts that never materialize must be managed proactively.

If you manage employees who are faking reality, make sure that you don't let this problem go on. A proper BI and execution framework will help identify fakes. Once determined, communicate how your view of reality differs from the view your employee is proclaiming. Design KPIs and choose metrics that represent the solution you're going to apply. Enforce execution until the problem is resolved. In this 2020s recession reality, you have no luxury to fake the facts.

State-of-emergency events like a recession make management and employees tell the truth. Success and failure are black and white. High performers will still make money while fake people will likely struggle to keep full-time employment or make any tangible success in their own businesses. When in dire straits, businesses can't afford to tolerate forecasts that never materialize or underperformance in a growing industry. In my experience, economic slowdowns bring the efficiency and accountability every company needs. Don't wait until then to make necessary improvements. Have a critical look at it today and start making your business recession proof.

Tip 2: Define What Is Fake

Advice to fake it till you make it often motivates many corporate employees to consider quitting their jobs to start a business. There is no convincing reason needed beyond that one coworker who believes in them. I do think that everyone needs to try to be their own boss at least once in their careers. That's a no-brainer. There are many beautiful, inspiring stories of people who made it. These stories are outliers. The 1 percent applies to any new venture an ex-employee starts. To join the shiny club of the 1 percent of entrepreneurs who succeed, you need to think and act differently from when you were successfully employed. Convincing other people to buy from your business is tough. Pitching investors and raising capital is tougher. And making a project successful is a different story. As an entrepreneur, you can't get away with false claims or naive plans, even short-term ones. First of all, you are in charge to pay all the expenses; second, any fake success like activities are quite visible from the outside. Everything is transparent in the digital world we're living in.

Many people confuse the fake-it-till-you-make-it business paradigm with simple negotiation tactics where you are trying to look better and bigger, set higher expectations, and raise the bar.

When I was twenty-five years old, a business partner and I started a trading company. We had a beautiful idea. We saw an opportunity in Eastern Europe, in a place where the market was underdeveloped. And we were just in time. All we had to do was convince a key supplier to bet on us, instead of on another company. We spun a story that showed the facts. I sold our six-month projections to our supplier as if we were already there. This clearly set high expectations, as we had zero revenue at that moment, but we had customers who had committed to start buying from us. I made nothing up. We had real revenue, and we were growing fast because we had acquired a customer base in advance. We had a clear understanding of the sales channel and its capacity. We knew a dozen customers by the decision maker's name and had preliminary agreements to trade. The supplier understood us and believed in our start-up's potential. We built a great business together without faking it.

Will convincing people to believe in you still work today? Yes. Communication and negotiation tactics help you never give up. If you get rejected, you can still believe in your business. You keep knocking on doors. You keep pitching as if you are already a successful business. You keep altering your offering, your sales model, and your communication strategy. To show confidence in what you do, show that what you say isn't imaginary but real. Just don't confuse this communication tactic with building a fake reality.

In corporate, using your track record to position yourself as successful is a common short-term success strategy that works for many people. Why short term? Because it buys you time. It gives you credit up front. It will not fix the fundamentals of your job, team, or business if they're broken. You still have to get it sorted. You can only make big promises if the foundation to deliver and a realistic execution framework are there. If not, you just bought yourself time to fix what's broken and start to deliver. Inspiring promises can't be accounted for in the organization's P&L.

Maxim, how can you tell me not to fake it till you make it when you lied your way into your first job after military school?

If you haven't wondered about this by now, you soon will, so let's resolve this apparent contradiction. I used negotiation tactics till I got my foot in the door. This worked because it was the beginning of my career. When you have no track record, no references, and no practical business experience, all you can do is convince people to give you a shot. Even then, I prepared the best I could. I researched the IT start-up, their industry, and the job description requirements. I knew enough about the role to have an intelligent conversation about logistics and get a job offer. In two weeks, they figured out that I knew nothing about logistics. The only reason they didn't fire me was because I had demonstrated other skills, and they saw my potential. I worked hard and committed to the tasks I was assigned until I finished them. Even though I was slower than they expected, I always got the job done. More importantly, I offered to go the extra mile to compensate for my shortcomings.

When you are pitching your first customers, your first investors, or your first partners, always broadcast the best-case scenario. This is fine if your well-calculated business model can survive the worst-case scenario.

Success is a self-fulfilling prophecy. When you get those first contracts, you no longer need to talk about what has been done in your previous roles and ventures. You start to talk about what *you have* accomplished for customers. That's the difference. That's why it works as a short-term tactic only. If you're growing a start-up, you're supposed to generate cash flow and be profitable. Otherwise there is no point in starting a business. I am not a big believer in taking five to seven years to build the foundation to keep venture capital coming into the company and still have no profit in place. You're doing everything you can to make money from year one, including broadcasting the best-case scenario to acquire first contracts. Otherwise you're wasting a lot of time, effort, and money risking no traction. During the last decade, there was so much money in the system that many new ventures didn't even worry about being profitable. Our new reality will reward efficient businesses that generate profits, not promise them in the distant future. Be clear about the difference between creating a fake business and using efficient short-term tactics.

Tip 3: Hire Slow, Fire Fast

The best way to avoid fake employees is not to hire them. This is easy to say but challenging to accomplish. Hiring is never easy, especially when you hire for leadership positions. In most cases, recruiting is designed to focus on what people have achieved, not on what they can achieve in the new role. The hiring manager is the one who is able to close this gap and identify the real potential of a candidate. To avoid a wrong hire, I recommend that you do the following:

- Hire slow. You are always pressed by deadlines. I get it. But don't rush to fill the role for the sake of a deadline if you don't have hero candidates to choose from.

- Don't waste your time on "tell me about yourself" questions. These won't help you hire the right person, as you still lack an understanding of what a candidate is capable of doing to help you meet your objectives.

- Filter the first impression. It's easy to hear what you want to hear when you're interviewing. Don't let enjoyable conversation stop you from being specific in what you are looking for from the leader you want to hire. Have a list of the one-, three-, six-, and twelve-month tasks and deliverables ready. Get specific answers about how every task will be accomplished, what challenges the candidate expects to face, and what resources are required to get things done. Make sure that you eliminate those who warm up the room by burning banknotes in the fireplace (i.e., cannot manage a lean business).

- Be critical about who the candidate knows or deals with. Understand the status of their relationships and their leverage. Otherwise it might look like Maxim Frolov knows the queen of England. I know her, but she doesn't know me. In other words, beware name-droppers who cannot make a deal.

- Take your time to get proper references—not the LinkedIn ones where you see positive things only. It's important to find out how the candidate made them happen. You should also use Glassdoor and other public resources to see if anything online conflicts with what you've heard. For example, a candidate may have a great résumé, an awesome LinkedIn profile, and credible references. And they tell you pretty much what you were expecting to hear. But then you find out that there was an exodus of employees at their previous employer because this candidate stirred up a toxic environment. Don't become an investigation freak, but pay attention to red flags.

- Don't expect a magic wand candidate. Divide what you have heard by ten. If that still matches your worst-case scenario in terms of the deliverables you are prepared for, you are safe to hire.

Everybody has a right to hire wrong. There are a million reasons why it happens. What we don't have a right to is to fire slowly when the hire can't deliver on their promises despite low-performance management. Firing is always unpleasant and never easy, especially when it comes to leadership positions. However, when you are clear that it's not going to take off, fire fast.

Don't fire fast emotionally. Fire based on unbiased data. Build a clear execution strategy for who will take on leadership duties once the person is fired. If it's you, make sure that you have the relevant time allocated, otherwise you run the risk of defaulting into seagull mode. Build a proper communication strategy to use in the event of displacement. See it as a team morale improvement opportunity, not as destruction. Be clear about how the team's daily routine will be impacted and address it proactively. Most importantly, don't procrastinate on firing. Do it yourself in the most ethical and professional way.

From about 2015 until early 2020, there was a strong employee market. The talent pool was disproportionate to open positions. In other words, it was difficult to hire a great professional for your budget. But it was easy for average candidates to get a great job. Stable economic growth brought unemployment rates to historic lows. That all changed thanks to the pandemic and the ensuing economic crises. In a recession, there are fewer jobs than candidates, which means you have a great chance of affording better professionals. Simply deploy my recruiting and vetting tactics. If you're looking to hire a leader, know what you need. Crisis management and the capability to run operations on the lean P&L are the most in-demand qualities for leadership positions. Outline the real skill set, abilities, and leadership style you need. Build a clear list of expected deliverables for the first twelve months in the role. That's how you get yourself ready to acquire the best talent and leverage the recession to build the best team possible.

CHAPTER

Fail Forward: Terrible Advice about Trying New Things (And What to Do Instead)

If I had to bet on it, I'd say that "fail forward" is the most common motivational phrase heard around the world today. Fail often, fail big, fail forward. On social media, failure is point A, wealth is point B. Motivational speakers and business gurus don't tell you how to get from one to the next, of course. In the real world, nobody aspires to fail. So let's stop pretending that failing big is somehow noble. It is not. But there are valuable lessons to be learned from our mistakes, lessons that can propel us toward success if understood and applied properly. What does this mean for young risk-takers? For people in their thirties, forties, fifties, and beyond?

We can all agree that failure is a rich source of learning. It's one of the fastest ways to get practical experience. This is an integral part of the human learning curve. It's how entrepreneurship works. But most cases of so-called failing forward today are profound wastes of time and money, especially when it comes to the popular start-up model inside corporations. Disrupting outdated business models, processes, and product portfolios is vital for companies to evolve. If they don't, someone else will disrupt their industry. Starting noncore businesses internally is the most efficient way for corporations to accelerate change. However, in most cases it's done with a corporate mind-set. As if the money to invest in the internal project is endless and hands-on leadership is not required. As if it's fine that the start-up loses money. As if everyone on the project has permission to fail without consequences. This is *not* how anyone learns anything.

The top 1 percent of the world's most successful people say that they're proud of their failures because those mistakes brought them to where they are today. But what exactly did they learn from their failures? How long was their journey to the top? How many steps did they take to get there? And how much longer was the journey than it needed to be because of failure? Failure is failure. It sucks. There's no reason to be proud of failure if you haven't turned its lessons into success. Failure for failure's sake only takes you backward. However, it *is* possible to turn failure into success. Here's how.

The Dead Horse Theory

The tribal wisdom of the Dakota Indians, passed down through the generations, says that when you discover you're riding a dead horse, you get off, bury it, and get a live horse. However, in business, other strategies are often employed to address the dead horse problem, such as the following:

- Buying a stronger whip

- Changing riders

- Threatening the horse with termination

- Appointing a committee to study the horse

- Arranging to visit other sites to see how they ride dead horses

- Lowering the standards so that dead horses can be included

- Reclassifying the dead horse as "living impaired"

- Hiring outside contractors to ride the dead horse

- Harnessing several dead horses together to increase speed

- Providing additional funding or training to increase the dead horse's performance

- Conducting a productivity study to see if lighter riders would improve the dead horse's performance

- Declaring that the dead horse carries a lower overhead and therefore detracts less from the bottom line than some other horses

- Rewriting the expected performance requirements for all horses

- Promoting the dead horse to a supervisory position

I saw almost all these dead horse maneuvers play out in one business in response to one crisis. It was September 2018. I was running Kaspersky's worldwide sales during their highest point of geopolitical pressure. In the United States, the Department of Homeland Security shook the company down for allegations of spying activities and connections to the Russian government. As you might know, every US government organization banned employees and contractors from using any Kaspersky product, even the free antivirus software.

As the person responsible for delivering on the company's revenue target worldwide, I was in a desperate situation. The United States is the largest market in the world. Losing anything there impacts you everywhere else. To understand the scale, just think about the $3 trillion gross domestic product (GDP) of California, which is bigger than the GDP of the United Kingdom,

France, and even India! New York City's GDP is bigger than the GDP of the entire country of Russia. Whatever sales targets we set for the US, our salespeople would never meet them. To compensate for losing just a few percent of our US revenue, we had to increase sales in every other territory by 5, 10, or 15 percent or more.

Geopolitical pressure turned out to be the small push that collapsed the house of cards. The real reason Kaspersky's US sales performance struggled like never before was not some conspiracy theory pushed by politicians and journalists. It was the evolution of the sales model. The transformation the company needed long ago wasn't happening fast enough. Geopolitical pressure was just an accelerator. In the US, Kaspersky started as a pure consumer player, a big presence in brick-and-mortar retail. Over time, the business evolved to offer direct-to-consumer digital sales, built a partner ecosystem, and developed small business and enterprise. Five lines of business in total. No one ever asked how each line of business contributed to profitability or how efficient the partner ecosystem was, as all up numbers were fine. Driven primarily by consumer business, North America's operations were OK, but the company was not set up for where the market was headed next.

This was the situation. We had big overhead, such as a massive headcount—employees who were hired during the good times when revenue justified more people. When revenue fell, payroll costs, fluffy marketing investments, and an unjustified travel budget killed company margins. The worst part was that no one could tell me what was going on. We had five lines of business. One of them contributed the most to revenue and margin and had the smallest team. What was wrong with the others? And what was the recovery plan? No one could say. Everybody blamed government inquiries and fake news. Those are big problems. Nothing good happens when any company in any country faces pressure from the local government. Of course that impacts sales. However, the best response to any external pressure in a competitive world is still execution excellence.

In December 2018, I made an out-of-the box decision. I took a second role as North America acting managing director. I firmly believe in unbiased decision-making. That wasn't possible in my current role with all the stories going around, morale declining, pressure mounting, and targets going unachieved. I needed to be objective so that I could diagnose the real problem and figure out an urgent reinvention of our sales model. I knew that I could do that better than anyone we could've hired because I brought synergy—I already had global experience and had gotten North American experience while walking in the managing director's shoes. Hiring someone else would have been a dumb move anyway. Think about it. When a company is in dire straits, who wants to join it as a managing director? Most likely not our dream candidates, who could make the difference we needed. I took the role myself and committed to doubling profit even as revenue declined. Most of the company's US business was annual recurring revenue contracts, which were melting. Customers refused to renew. We weren't growing new business because many customers, especially in the B2B space, were either spooked by headlines or restricted from buying anything from us.

The largest American retailer, Best Buy, had terminated Kaspersky's contract in September 2017. This loss had no chance of being compensated for by the few minor retailers who were still working with us. Direct-to-consumer sales were demonstrating solid performance but had limited growth potential to compensate for the loss. Because North America was heavily dependent on consumer sales, the only way was to ramp up B2B, which had been bleeding for a long while. Our competitors smelled that blood in the water. All they had to do was pick up the phone and ask our customers, "Have you seen the headlines? Are you sure you want to continue with a company like this?" Good for them. If I were them, I would do exactly the same. We all would.

One of the options was to shut down the office and focus on what we could manage remotely. After all, news was going from bad to worse. However, I convinced management to take a risk. To turn North America's business around and, as a result, make the entire company sustain the required performance, I committed 100 percent to the managing director responsibilities.

I started with a quick internal assessment. *Who does what?* We had five lines of business and several people supporting each line. Similar numbers, even though revenue per line was disproportionate. After my unbiased, cold hat review of organizational structure, business processes, and key metrics like revenue per head and sales coverage, I came to a conclusion. Believe it or not, we were overstaffed. My report stirred up controversy. How could we have too many employees when we needed more people in the field selling stuff?

In reality, more feet on the ground wouldn't help in those circumstances. The bottleneck wasn't the number of people dealing with the problem or doing sales. The business model was extremely inefficient and had no chance to survive the crisis. Why? The team looking after the small business customers was doing nothing, but the direct salespeople with their low ADS blocked their attempts to scale up. The team looking after the large enterprise customers were doing broad sales and marketing activities, which eliminated their chances of running a proper sales cycle and closing complex deals. The sales models were turned upside down with questionable deliverables and sustainability during the crisis time. The only way for the team to try to make their numbers was to sell to a larger group of customers at the same low-ticket price. Not the most sustainable model when a storm hits the shore. They were stuck. They had no time to build relationships with customers because they had to cover too many accounts. The only choice we had was to push, push, push. In sales meetings, our people had to get straight to the proposal, ignoring what customers really wanted. A straightforward push sales method. Like a Mike Tyson punch. Just much less efficient.

Acting under constant underperformance pressure, the North American team adopted fail forward as their business culture. The majority of sales, marketing, and business development strategies and tactics were based on terrible thinking, such as "This is the best we can do in this awful situation," "It is better than nothing," and "It's always worked, so let's do it more intensively." These fail-forward practices got the team stuck in a loop they couldn't escape from.

Understanding what our product portfolio has to offer and building a solution to the enterprise customers was our best (and only) alternative to catch up on revenue. Despite all the gaps, there were several things that were more than sellable. However, the team had never tried to sell them before. When you're selling complicated, comprehensive solutions like enterprise cybersecurity, the customer must trust you. The fastest way to build this trust is through the consulting you offer the customer. It's the trusted adviser model, and it's free. Patiently discover the customer's business processes, objectives, and agenda. Educate them on the latest trends and threats. Advise honestly. Build your credibility. You have to build this relationship on the deep knowledge of what you sell. It's very different from being a pushy salesperson who squeezes customers into corners until they buy. As you maintain trust, you preserve that relationship and build a solid base for future upsell and cross-sell opportunities. Basically, you can increase your deal value with the customer every year.

At Kaspersky US and Canada, no such business model was in place—yet. Low office morale and "this is my last day here" letters didn't help leaders step back, see things in perspective, and figure a way out of the crisis. The massive office in Boston got quite empty, which impacted morale even more. If you're trapped in a situation where customers and partners terminate agreements every day and you know you're not going to make your target anyway, you get stuck simply doing what you've always done—just more intensely.

I reported all this to the board of directors. My findings, my conclusions, and my vision surprised them. Since early 2017, when the company faced problems with the US government, they had based their decisions on presumed loyalty, not on business model efficiency or team shape. That's not what helps improve an extreme situation.

While crisis-managing the region's P&L, I applied strategic operational changes to each revenue-generating part of the business and each quota-bearing team. I had to stop retail business in the US (as it wasn't profitable), I reshaped small and medium business (SMB) and enterprise teams to get profitable again, and I changed the way marketing investments were made. The way-out-of-crisis plan was based on three fundamental things: business efficiency, personal accountability, and transparency. The best way you can address low morale in huge empty offices is to openly show why the business model that was there for decades doesn't work anymore, then share simple-to-execute granular business plans to carry them out of a desperate situation. Of course, twelve-month anticrisis plans had to be well prepared in advance. Layoffs, organization and leadership changes, and sales model changes were an integral part of it. Changes in leadership were crucial to support this crisis management task.

While charting the way-out-of-crisis plan that presumed the worst-case scenario of declining sales, I decided to experiment and start a line of business this region never had—strategic enterprise sales. A line of business that brings ten times higher ADSs and targets larger customers than this team used to deal with. This line of business was already fast growing in the Middle East, Africa, and Russia but not yet in the US or Canada. We had nothing to lose by trying, even in the situation of melting sales revenue and customer churn. If it worked, we would shorten the path out of the deep crisis. Worst-case scenario, we would downsize the team and expenses to preserve the required business margins.

I explained my plan to the vice president of enterprise sales of North America, Rob Cataldo, that we were going to start something that his salespeople hadn't tried before. I proposed that we build a solution-selling practice and sell complex solutions based on the most intelligent stuff the company had to offer. This was a paradigm shift for the team, as this new sales model presumed that they would sell to decision makers a few levels above where they used to sell. We understood that not every salesperson was ready for this in terms of both hard and soft skills. So we assumed that we would build it from scratch as a pure start-up. The plan presumed that geopolitical pressure would get worse, that Rob's team wouldn't be able to bring in any new logos at all, and that the existing customer pool would continue to churn at the same pace as the last twelve months. We would build a P&L based on this assumption. To match this worst-case scenario, we'd downsize Rob's team three times over. Simply put, maximum revenue from minimum personnel.

Rob agreed to my plan. The first step to execute was a market mapping exercise. The idea was to change the way renewal contracts were handled. Instead of approaching customers based on the renewal date, we decided to do it based on deal value segmentation, with largest deals first. It didn't matter if the renewal date was next week or in six months. Then midsize deals. The idea was to get clarity on what customers would not renew for sure and write off those deals from the sales forecast. Those who stayed were incentivized for upsells and early renewals. All to secure the worst-case scenario deliverables.

All small enterprise deals were moved to the SMB team. Yes, we ruined existing sales taxonomy. But the idea was to figure out our worst-case scenario sales projections to free up the sales team to deploy our all-new sales model—solution selling. Because the enterprise team was downsized by two-thirds, we had to change geographic sales coverage. The new, smaller team had enough bandwidth to cover bigger territories and more accounts than they used to, since we moved all their smaller accounts to the SMB. We also involved a loyalty index—a metric to measure customer satisfaction and loyalty—to mark the renewals we were likely to lose. For example, were customers happy with the products they purchased? Were they happy with customer support and account management? Based on the number of deals of each type, we were able to calculate the expected workload per account manager. Finally, the team had a shot to increase profitability in a declining revenue situation.

We visualized the plan; set the road map, specific metrics, and tracking methods; designed necessary trainings; and converted performance reviews into workshops where every new deal went through vetting for errors, strategy, and next steps. To build confidence, I led some of the trainings, making sure that I transferred my experience and knowledge directly to the future enterprise sales champions. No well-polished training provided by human resources can match the efficiency of the coaching that sales leaders do. Sales leaders can help identify weak links and give fast feedback. It helps to adjust to the local specifics. It helps to ignite salespeople, as they always respect the time and effort invested.

"We're going to evolve with the times," I told the sales team. "You're struggling because it's easy to replicate what made you successful a long while ago. We can't expect that it will make us successful again. 'The problem isn't me,' I've heard some of you say. 'This geopolitical situation impacts sales.' Not really. Our jobs depended on building a new line of business. Excuses and entitlement won't cut it."

I sensed that this entitlement also explained the team's hesitancy to run with the new plan. When you work in an organization under serious pressure, a lot of your colleagues escape your toxic employer to find a better job. Everybody's got to pay their bills. If you stay during these circumstances, you feel a certain entitlement. You think your employer owes you because you showed loyalty. Trust me, they do not.

The bigger problem I observed was the fear of failure. The team didn't want to do something they'd never done before because of their fear of failure. That explained why they were struggling to find new prospects. It was easier to keep giving excuses about why the customer base was melting because of geopolitics than it was to step out of their comfort zones and sell in a way they'd never had to before in their ten-year-plus careers. Execution excellence would build their confidence and cure their fear of failure. What happens when your team's fear originates from constant underperformance? People panic, do a bunch of chaotic things, and expect something to work. This is not what helps in crisis situations. Fear saves, panic kills. Decisions and activities must be based on well-analyzed numbers and executed with a purpose. Unfortunately, the purpose behind most of these chaotic actions is being able to use "we did our best" excuses.

During the first six months of executing the plan, the US and Canada team made a remarkable breakthrough. We reshaped the organization's structure and business processes, changed the market positioning, and improved team morale to the level where attrition has been back to normal. Across North America, the US team included, enterprise salespeople made progress in transitioning from commodity sellers to comprehensive solution sellers. They moved to the higher league. They closed deals bigger than they ever had before. Using their new skill set and sales process, they were able to acquire new customers from the world's top ten "dream clients" list. All with the government ban, geopolitical pressure, and an incomplete product portfolio remaining.

We went through painful layoffs and massive changes during that short period. This transformation bore new leaders. While applying changes to the SMB sales, I bet on Matt Courchesne, a young SMB guy with big potential, giving him the opportunity to manage all the US SMB business. He delivered above expectations, making the transformation ball roll. Enterprise Vice President Rob Cataldo and his team made a huge difference while building this new line of business. The way Rob pumped up his leadership and management skills was truly inspiring to the entire organization and set a new bar for all sellers.

The fastest way to build muscles is to work out. The more intensive the exercises are, the faster you evolve. Twelve months after we committed to the US business transformation,

Rob was promoted to managing director of North America. Despite the negative press and toxic public perception, the team was able to improve the renewal rate by upselling value-added services, preserving more customer relationships than expected, and demonstrating impressive enterprise sales revenue growth. North America's business remained in the middle of the geopolitical storm, but sales became predictable, and P&L improved big time. The business got a better base on which to operate and keep evolving in this complicated environment.

What made launching a start-up inside the corporation in the middle of a deep crisis successful? Do the right thing, and you get the right result, correct? It wasn't that simple. I'm not talking about leaders' or employees' resistance or gaps here. I'm talking about the fail-forward paradigm we had to fix to address crisis circumstances. Excuses, entitlement, and incompetence were symptoms of the broken business model, but they were not the root cause. The root cause was nonstop failing forward during crisis management—the absolute worst time to fail.

Tip 1: Fail the Micro to Progress the Macro

Fail-forward advice misleads many people simply because there is very little clarity on definition. If we separate failure into macro and micro, we get clear. Failure on a macro level ruins countries, companies, and careers. Waste your company's time or money failing forward, and it's game over. Failing-macro advice can come in different shapes and forms. Bet on just one deal to make your sales number, keep investing aggressively to get this one unicorn deal, fail until your lucky number wins, or keep failing and expect that this will bring you a win in some magical way. Not everyone is ready to go all in on just one big idea.

Statistics confirm that success driven by this is an exception, not a sustainable business model. Failing macro will likely not take you far.

Micro failure is a different animal. Micro failure is when you take risks, try new ways of doing things, and test new product positioning, new marketing concepts, and new sales models, but all on the micro level. Break down every task or project to its micro parts and allow failure within one small part only, all with no risk of changing the direction in which you are heading. Set the process for how you'll monitor and identify failure's root causes, learn from them, and rework. All based on the data and on limited time frames. Micro failures can make you successful as long as you allow them to show you where your next step is and where you need to build competence.

If you are in sales, you have a little room for fail-big/fail-forward experiments. Your results are part of the company's budget and P&L. These numbers go in your bonus calculations and performance review. You might have learned something from failure, but a thousand failures earns no one a promotion. Only results do. Micro-failure methodology helps to get the results you are looking for.

When I proposed to launch a new line of business and target the largest potential customers in the country where the company was banned by the government, I was clear about our room for experimenting—that is, micro failing. We built it based on the worst-case scenario P&L, which enabled us to secure the required level of profitability.

Once we secured the required level of recurring revenue to meet P&L projections, enterprise sales managers started to apply the knowledge we transferred to real life. Failure after failure. But there was no punishment for it from management, as it was all preplanned. Instead, every case was reviewed, errors were corrected, and messages were altered and then back we went to the field to try it again. We agreed that we would manage the travel and entertainment budget super lean. We would travel only when there were circumstances to move enterprise deals to the next stage. We made sure that there would be no endless entertainment of customers with no specifics about what the customer's business goals, objectives, budgets, and time frames were (e.g., what exactly we can sell to this customer?). The most important micro-failure tactic was to kill the fear of failure, to get everyone used to failure on the micro level to progress the macro and to help them grow their competence and results through their chain of micro failures.

Learn from your micro failures and those of your employees. Through these micro failures, everyone can learn what does and doesn't work, test new things, and drive constant progress. Every successful start-up does this, especially start-ups inside a company, a phenomenon known as intrapreneurship.

You probably know that Amazon Web Services, the world's largest cloud storage provider, was a start-up within Amazon itself. Nurtured within an environment that allowed micro failure to prevent macro mistakes, AWS, as it's known, had more than 30 percent of the cloud market in 2019 and demonstrated more than 30 percent growth.[16] That's more than the three competitors combined, Microsoft, Alibaba Cloud, and Google Cloud. What made AWS such a success was Amazon's own need for it to work. They could not risk pointless failures, missed timelines, and overblown budgets, all problems typical of failed start-ups. Why? Because Amazon was itself the first AWS user! The web services

---

[16] Ron Miller, "How AWS Came to Be." TechCrunch. July 2, 2016. https://techcrunch.com/2016/07/02/andy-jassys-brief-history-of-the-genesis-of-aws.

company dates back to the year 2000 when the young ecommerce company needed scalable storage to manage hypergrowth. In the words of Amazon Web Services CEO Andy Jassy, AWS started "out of need." Because Amazon's ecommerce business model already had low margins, Andy and his people had to work as lean and efficiently as possible. They could not take unjustified risks or throw money at problems. AWS did not exist in a vacuum but was developed, tested, and iterated by and for Amazon every single day. Once AWS had proven itself to Amazon's developers and third-party merchant partners, it was ready to stand on its own. The industry leaders in the cloud space at that time had no clue what hit them. The rest is history.

To create an AWS-like learning framework for yourself at your job is hard. But it's vital to have. The faster you can expose an idea to the real world, the sooner you'll get feedback. Micro testing and micro failure can bring quick lessons that turn into success. If you're in a management role, it's your job to set up such a framework that makes employees successful. That success comes from dealing with micro failures like a bug, a break, or a question that no one knows the answer to (yet). Challenge people to fail on a micro level—not for the sake of failure but for *risking* failure—and convert failure into next steps.

In short, micro failures are a rich source of progress. Companies like Amazon proved this. Faced with a real problem, the need to scale but without the infrastructure to do so, Amazon assigned a team to try different things. Within three years, the little AWS experiment worked.

Micro failure helps you to stay on course, learn, and adjust your plan, your strategy, and your execution. Fail micro and fail often if you want consistency in your progress.

Tip 2: Watch for Red Flags

While managing micro failures, watch for red flags. When I helped Kaspersky North America transform their broken sales setup into a competitive, profitable business despite the government ban, the number one red flag I saw was that "We've always done it this way. This is just how it works in the United States." How silly. "This is how it works in the US." "This is how it works in India." "This is how it works in France." Every geography has its own specifics, dynamics, and business culture and its own way people collaborate, negotiate, and get things done. But sales are sales. Profit is profit. Loss is loss. Two plus two equals four in every single country in the world.

If your sales model doesn't deliver the expected outcomes, you have no choice but to challenge it. Find what is broken. If it's not relevant to the market, then build a new model that works. Dismount the dead horse.

Efficient crisis management always starts from understanding bottlenecks. Only by knowing them will you be able to change direction. To identify them, watch out for red flags. In Kaspersky's particular case, we looked at the following:

- Average deal size

- Deal velocity

- Win rate

- Customer churn

- Renewal rate

- New logo/recurring revenue split

- Sales coverage (number of customers per seller)

Once we moved to execute the new strategy, we set a task-specific red flags watch list:

- Pipeline value and coverage dynamics

- Pipeline quality and balance to avoid having just a few large deals covering the necessary sales numbers

- Number of new logos in the pipeline

- Number and value of new "solution" deals

- Deal velocity progress (i.e., how many deals moved to the next stage)

- Renewal rate dynamics and percentage of deals upsold

- Win rate dynamics

It's critical to stay committed to monitoring red flags. In most cases, the sales manager doesn't see it the same way. When you do a lot of customer interactions and sales activities, you have a strong feeling of constant progress. Ignore the feelings. Base your conclusions on numbers and unbiased assumptions only. This will help you to better assess risks and opportunities.

While incentivizing the US team to try new things, one of the directors suggested a white-labeling initiative. White labeling with the largest software vendors and telecommunications companies (telcos) was one of the oldest lines of business for the company globally. The difference this time would be the sales channel. The model presumed that we would use telcos' massive consultancy channels to reach customers we never had before. It was a new twist to address small business customers. This sounded like a real opportunity to create a great revenue stream without facing geopolitical pressure. No Kaspersky logo, no problem.

Hype was created, and the team spent a lot of effort, time, and money on travel, entertainment, and engineering, but the results weren't there. While reviewing why the white-label model was not taking off, I figured out that the project was designed to hardly break even. The director who proposed and supervised this initiative didn't manage the project as a start-up expecting to make a difference. He managed it as another corporate push, treating money as an endless resource and bearing no responsibility for results. This was a classic macro-fail scenario for the sake of showing an attempt to do something in a desperate situation. This does not help to manage your business out of desperate circumstances.

I recommend setting your own methodology of red flag monitoring based on key metrics and the KPIs representing the expected outcome. Identify the need for error correction earlier and help your team save time and effort, achieving the result faster. Lead it personally. Don't expect every stakeholder to do their part with 100 percent excellence and accuracy. Watch your red flags. This will help to keep failures micro.

Tip 3: Fix before Firing

Failing forward and failure itself go hand in hand with low performance. Every organization has P&L objectives. Profits and losses. You need the right balance. If you or your employees fail, your P&L formula breaks. Profits down, losses up.

What do you do when this happens? What if, in your situation, people keep failing forward without recovering yet show no signs of recovery? It's time to deal with the low performance. In more than twenty years in the technology industry, I've found that most people do not know how to manage low performance—it's tough.

It's easy to just say, "You're fired." The real world doesn't work that way. There are costs. Paying severances. Finding candidates. Interviewing them. Hiring and onboarding the right one. Then there are the time lags associated with firing an employee. Their projects and tasks come to a standstill while you figure out who's going to pick up where they left off.

To deal with the low-performance employee who keeps failing forward, you cannot simply say, "Failure is not acceptable. I cannot allow this underperformance to go on." This will take you nowhere. What will help the failing employee turn their job around is to assist them in seeing what is not working and how to address it. Otherwise, it is all perceived as unhealthy pressure. This moves nothing forward. Consistent enforcement is the key to address low performance. Don't let it go and expect that it will improve by itself. Here's how:

1. Identify the root cause of underperformance. Is it an incapable employee, a bad business value proposition, poor product quality, or market irrelevance? In rare cases, the cause may be negative market dynamics or so-called force majeure situations.

2. Ask the low performer to come up with a recovery plan. Set the right expectations. You need to get a granular, numbers-based, easy-to-execute proposal.

3. Once you receive the proposal, analyze it and challenge its realism in terms of time and resources required. Alter the plan as needed and discuss it with the employee.

4. Set and enforce weekly or biweekly performance tracking. Don't set control points too frequently. It's hard to measure progress quicker than in one week. Correct errors and guide the employee in addressing the low performance.

5. Provide clear feedback about the low performer's progress. Help more than press. However, pressure to create a state of emergency is a must-have.

6. If progress is not satisfactory, get ready to say goodbye and replace the employee with a better performer. Be objective and willing to fire even people you like and have known for a long while.

7. Design and execute plan B to compensate for performance gaps. Execute it from day one of the low performer's recovery exercise. See what your top performers can do. Can they overdeliver? Align important incentives for them to make success happen.

8. Don't be naive and expect a low performer to tell you the ugly truth. Don't buy bullshit like "Don't worry, we'll be fine by the end of the quarter." Don't become the victim of your desire to hear only good news. You need to operate with precise information to give yourself a chance to catch up on the performance.

Low-performance management requires criticism. Shake and wake people up. All precise numbers are unbiased. Run practical coaching sessions on how to do the job right. Raise the bar, then help people meet it. Higher expectations and practical help changed everything for Kaspersky North America. What no salesperson realized was that a $1 million deal requires the same time and effort as a $100,000 deal. The one difference is confidence based on competence. Do you have the backbone to deal with the bigger number? Have you built the skills to do that? Are you ready to take responsibility if a $1 million prospect says no and the deal disappears from your forecast? Do you have a plan B to compensate for it?

Poor performance management does worse than the opposite. Bad managers compound failure on top of failure until nobody has any idea how to fix that broken P&L formula. What do you do when you realize that an underperforming employee is hopeless? That your only recourse is to let them go? That you have a better chance of success if you hire a replacement? Believe it or not, many managers would prefer to do nothing. Experience has shown me that successful managers take responsibility for employees' lack of performance—and lack of potential. If you are not capable of firing people the right way, you are likely not building a micro-failure framework to help employees improve. You can make mistakes while hiring, but you're not allowed to make mistakes in firing.

Tip 4: Performance-Manage Your Way to Success

What if your low performers refuse to change? Fire at will. Once you spot underperformance, you cannot take it easy and leave it unresolved. The problem will keep growing and never go away by itself. I had a few pretty bad performers at Kaspersky that time. They're nice people. We're still friendly. But business is business. When you see an employee incapable of handling complicated tasks, you've got to split. It will not get better. Remember the mantra of every salesperson: you're only as good as your last sales quarter results. If quarter by quarter results are not good, if you see excuses but don't see goodwill or practical improvements, make the decision. Low-performance management is an integral part of company culture. All people managers work hard to create a unique atmosphere that makes employees happy to wake up and commute to work every day. Of course, relationships always go beyond boss and employee. You need trust. Even friendship. Yet to be ready to fire a friend who cannot perform is the must-have skill for any manager who wants to be successful in the long term. That's the way it is. Saving people from the consequences of underperformance because you have sympathy dooms both of you, your team, and the entire company.

I'm not saying that firing people should make you happy. It's an unpleasant situation for everyone. Most managers aren't willing to make themselves uncomfortable, even if it's the only way out of a tough situation. Such people would rather point fingers than manage low performance. It takes no courage to seagull all over your employees. "I cannot accept your low performance." "I want you to work harder." "Next time show me the numbers." It's emotional, not helpful. If you're going to performance-manage your way to success, you need self-confidence and you need a plan.

Before I made big personnel changes at Kaspersky North America, we visualized our sales model. On one screen, on one piece of paper, we could see how every single account and business partnership was covered. It was no surprise that we had too many people for the size of our business. This simple mapping exercise separated essential personnel from the fireables. I recommend that you visualize your organization and its objectives and processes. Be clear about how your business or team works once you pull the trigger. There is nothing wrong with understanding that you are overstaffed; you simply need to find a way to redeploy resources (i.e., redundant employees' salaries). Organizations start from scratch, manage lean until success arrives, and then expand. At some point, expansion goes out of control. When it does, go lean again. It's the organizational life cycle. I did this exercise in the Middle East and Africa back in 2016. That's when I had to fire almost half of my salespeople and hire better ones. But first I was clear on what share of their responsibility my GMs and I would cover. We visualized the entire process so that we could *succeed* forward.

Once committed to low-performance management, you've got to stick with it. Either help employees improve or let them go. While firing employees, recalculate your entire business model to avoid the situation where someone gets let go, no one knows how to do their job, and you're worse off than before you fired that person.

Performance management isn't easy. Nor is performance itself. Fortunately, these tips will help you get people aligned, inspire peak performance, and help your company weather any storm. And storms will come, one way or another. Yours could be a public relations nightmare like Kaspersky's, or perhaps we'll all be in the same sinking boat when the recession slams the global economy. Whatever happens, only those who live in the real world are going to make it. I'd like you to be one of them.

Conclusion:

Eliminating Terrible Advice from an Entire Company

As businesspeople, it's critical that we bring the right ideas into our organizations. Just as terrible advice distributed across a business can produce disastrous results, so can an organization succeed when it's full of people who bring bright ideas into the real world. How do you get there? Simple. Consider yourself an adviser to the people you work with—even if you're not technically above them in the organizational hierarchy. Advising is different from bossing around. Advising is an extra responsibility.

Anyone who has ever been in the corporate world knows that many managers don't know how to advise. If they don't offer advice responsibly, they likely don't take responsibility for their people's performance and they don't develop their employees into top performers. Instead, they often become commanders. "Your pipeline is shit," they tell you. "We have three weeks to go until the new year. Bring me results." Fantastic. How are people supposed to do that? They're looking for practical guidance, not commands. This is what leaders are for.

Of all the advice I've given you in this book that I wish for you to start deploying, the most important is to be practical. Build your strategy based on your ability to execute, not on your hopes and dreams driven by hype or a desire to be successful. Offer clear, actionable instructions and a granular, simple-to-execute plan, not just generic motivational rants. Albert Einstein once said, "Everything should be made as simple as possible, but not simpler." Often, leadership issues vague mandates without any directions beyond "Just get it done." Such a simplistic demand is usually accompanied by an even simpler plan that ignores practicality.

Leaders frequently forget that employees interpret advice and ideas differently from them. Many leaders simply announce a shiny new goal and expect people to figure out how to get there on their own. The same happens in low-performance management when an employee's performance is not where it could be. Saying, "This is not acceptable" or "It's a wake-up call" won't get you anywhere. People can't figure out what they need to do unless you offer them guidance and advice.

It's the same for small business entrepreneurs. When people start a business with more than two stakeholders (e.g., cofounders), people often assume that crucial work will be done by someone else—despite the fact that job responsibilities are well outlined. The job won't get done by the mysterious Mr. Somebody. You feel like you've covered everything, but there's always either a blind spot or an overlap. Overlaps happen more often than blind spots because partners think it's their job to take care of everything. This wastes time and provokes frustration.

Success at the organizational level requires commitment to getting your hands dirty. Everyone needs to know the entire workflow end to end without assumptions, understand their role in it, and execute accordingly. Some love to call this micromanagement. I call it ownership and transparency. True micromanagement is a very different thing.

Let's think for a second about how micromanagement happens. What's the path to micromanagement? It's always a side effect of gaps in ownership. Usually there's no alignment, no control, and no mutual understanding of the direction the business is going. Maybe red flags and fires are everywhere. Management, having no idea what's going on and how to fix it, starts to micromanage. In most cases, we see seagull management hand in hand with micromanagement. When you don't own how your business processes work or you don't know your numbers, then you don't know if you're winning or losing. If you can't tell what's going on in the business in real time, you have no choice but to micromanage when things go wrong.

The simplest way to avoid micromanagement is to deploy proper numbers- and data-based decision-making processes. If you're in sales, you know how your sales target breaks down by every salesperson, territory, business segment, business partner, customer, and product category. You're able to dissect your target by the crucial metrics of efficiency. You're clear about how you're going to make your target. You're also clear about how to properly address risks and opportunities. It's so simple, yet too many sales leaders still stick to a mix of hopes and dreams. Data ownership and transparency are still the fastest way to avoid micromanagement in any business.

It's easy to build data management tools today, even in start-ups and small companies. Until recently, only the largest corporations could afford such tools. However, having a good look at your numbers doesn't help until you personally set data points that describe and represent your business through those numbers. Once you can read them properly, you can translate them properly. Once translated, you're able to execute your goals with excellence.

If a team member shirks responsibility for outcomes or balks at instituting measures to track their output, that person doesn't belong on your team. It's easy to assess employees' performance without interrupting their work. You can set up tools that provide you with unbiased data about specific parts of your business.

People tend to do things exactly the way they always have, so this might give you the opportunity to find a simpler process.

In the corporate world, admitting a problem can make you afraid of looking bad in front of management and peers. It also requires that a solution be brought to the table. Solutions can get ugly—firing people, cutting costs, or assigning extra work for the same wage. However, deploying self-awareness allows you to openly admit missed shots, take responsibility, and focus on recovery at every level of the organization. Don't let performance problems get to that point if you can help it. In recession circumstances, it's vital to act quickly and to find opportunities to reinvest instead of simply cutting costs. Be a leader whose perspective is granular. See every process step-by-step and demand the same from everyone on your team. This will make you a force of nature. You may be accused of being too demanding or too tough. But you're just doing your job and holding others accountable for doing theirs.

Each and every organization on earth today is heavily influenced by public advice—especially advice that social media helps distribute. All of us consume content that our newsfeeds give us every single day. The wisdom of crowds has an influence on business decision-making like never before. Such advice looks like it's written in stone because it's everywhere. But being irresponsible and following the wrong advice across an organization can damage results, both in the short term and in the long term.

How can you tell good advice from bad? How do you interpret the right advice for your situation and deploy it throughout your business? This is becoming harder due to the ever-increasing speed of life. Many corporations struggle to adapt to the pace of our fast-changing reality, consumption model evolution, and rapid shifts in consumer behaviors. Too many historically successful companies are living in the past, following the supposedly timeless advice they keep hearing from the same experts and expecting it to work in the new reality.

Many leaders run their business like it's 2010. The role of management has changed a lot over the past decade. Historically, the role of management was to maintain the state of business and to execute the rules, product calls, and processes written for the organization—processes written based on what used to work in the past. Decline in performance always correlates with a decline in employee productivity. Deadlines are moving, and many tasks require reworking. Management rarely cares about engagement, empowerment, enablement, or any aspect of employee development. Coaching and mentoring rarely happen at the required scale. Managers are advised to push employees to do what is expected rather than to do what makes sense. Reality has broken these practices. In the epicenter of the 2020 recession, we do not have the luxury of living in a time warp, repeating what made us successful ten years ago.

When you command and demand, you force people to make more calls and hold more meetings. In a rush to close gaps in performance, they tend to do everything at once in the hope that something will work. They might start to offer bigger discounts that devastate your margins long term. Then they might start to run several conflicting marketing campaigns at the same time, resulting in confused customers and partners and ultimately unachieved sales targets. When that happens, low performers are punished, and leaders have to dive deep into the root cause of the underperformance by identifying the problems, altering business processes, educating and equipping their team with necessary skill sets, motivating their team with a simple-to-execute plan, and starting out on target performance from month one of the next fiscal period.

So many leaders used to say, "We'll catch up by the end of the quarter." That's hope. Hope is not a plan. Good luck is not a solid business paradigm. When you read your numbers correctly and identify the root cause of the problems and opportunities, the next thing is to get clarity on *how* to proceed. Who does what? How does it match relevant employees' skills and the time required? How does it match the budget available and expected returns on investment based on a worst-case scenario? Deploy processes that are easy to execute. Be aware of problems—don't sweep them under the rug. When there's a problem, work with your team to resolve it instead of punishing them. Punishment is easy but rarely productive.

The 2020 coronavirus pandemic and ensuing economic disruption have made these common business problems impossible to ignore. When global supply chains broke without notice,[17] even supposedly stable industries like high tech, law, and professional services started to report layoffs, pay cuts, and furloughs.[18] In a situation where only essential personnel are allowed to work—and therefore earn—managers and employers alike wanted to be counted among them. Who do you think made the cut? The people who trusted good luck or those who had already proven that the company could count on their solid performance in the crisis environment?

---

[17] Avi Salzman, "Coronavirus Is Disrupting Supply Chains. These Industries Are Most Vulnerable," Barron's, February 29, 2020, www.barrons.com/articles/tech-apparel-profits-could-be-hardest-hit-by-supply-chain-shocks-51582938634.

[18] Taylor Telford and Jena McGregor, "The Latest Sign the Economic Downturn Is Intensifying: White-Collar Workers Are Being Laid Off Now," Washington Post, May 28, 2020, www.washingtonpost.com/business/2020/03/28/white-collar-coronavirus-recession.

If you found yourself leading a team or organization during COVID-19, you had to make tough decisions. Only people who could deliver the required results in a falling market got to stay. And by *stay*, I mean work remotely. Isolation changed both corporate processes and consumer behavior. In the wake of the virus, online transactions increased by 52 percent in a three-week period at the start of quarantine in the United States alone.[19] But one benefit of working from home independent of public safety is productivity. Remote workers are 24 percent more productive and satisfied than their cubicle-confined colleagues.[20] Efficiency is a must-have for any business enduring turbulent times, and you've got to lead it.

You probably learned a lot about your direct reports (and yourself) during that season, didn't you? For example, simply commanding people or expecting them to figure things out themselves does not help you stay competitive. Not only did you find yourself in a coaching role but also you probably needed a mentor yourself to navigate the turmoil more efficiently. I believe that coaching capabilities are a must-have for any leader today. I mentor three to four people a year on average, and I can say that I *get* an incredible value for my own professional growth from these coaching exercises. It's like living multiple business lives at once. When leaders ask me for advice on how to grow, I always recommend becoming a mentor. It will help you drive change into your daily business—and help you answer your own questions. Prioritize time for it and stay committed. It pays off.

---

[19] Mike Cassidy, "The Coronavirus Will Change Commerce and Consumer Behavior," Signifyd, March 12, 2020, www.signifyd.com/blog/2020/03/12/coronavirus-changing-commerce.

[20] Yi Jen, "How Remote Working Helps Us Become a Productive and Happy Team," Labiotech.edu, November 28, 2019, www.labiotech.eu/inside-labiotech/why-remote-working-is-helping-us-become-a-productive-team.

Coaching and mentoring are keys to making employees, teams, and entire companies wildly successful in the long term. Mentoring helps employees stay realistic and pragmatic, live in the real world, respond to new challenges every day, avoid anything-is-easy-on-social-media traps, put things in perspective, understand how business processes are connected, and adopt a mind-set of wanting to make a difference.

When Terrible Advice Takes Over a Company

Advice is a powerful tool. It enables, empowers, and fixes all at the same time. It's easy to ruin everything when the advice is wrong. In many cases, public advisers try to help, but their tips or instructions are too generic. They're not granular or practical. Sometimes they're not even aligned with reality. When taken seriously, they can mess everything up.

To offer proper advice that makes a difference and eliminates terrible ideas from your organization, do one simple thing: Deploy processes where transparency and communication of data and goals are based on the *how.* How will you make it happen? Until the *how* is clear, work with your team to find a solution.

Does anyone believe that general advice and motivational quotes like "Just do things right" will help move the company forward, close performance gaps, overachieve the fiscal quarter target, stop customer churn, accelerate new customer acquisition, stop wasting marketing budgets, or bring efficiency into the business? That's naive. But there *is* a simple way out. Be practical. Walk in the shoes of those who are supposed to execute the plan you're about to build. Anyone can have ideas, but few can execute.

The more specific you are in your communications and the more precise you are in what you're looking for from your team, the more influence you will have. People are busy. Most of us are locked into our everyday activities. Often it's hard to see the big picture. We're too busy to see opportunities and to notice the red flags in our daily operations. In the fast-changing reality we live in, business productivity starts from practicality.

Make practicality the cornerstone of your business. Challenge any decision you make with practicality. New product or service development, expansion to other geographies, mergers and acquisitions, hiring and firing, changes in the way you market and sell—all these must be properly criticized in terms of how and who will get it done, how long it will take, and what the worst-case scenario looks like. Incentivize ideas, but build the framework based on the *how*. Have fun executing your framework daily. Eliminate politics, no matter what size your team is. No business has the luxury of allowing them anymore.

Take full responsibility and put in the work. Many things will not go according to plan. Being ready to respond, finding solutions quickly, and keeping the organization competitive defines you as a leader—a leader who carries the team through turbulent times and builds a base of long-term success.

Be granular. Be practical. Lead by example. Only then will you sniff out and destroy all the terrible advice.

About the Author

Maxim Frolov is an internationally renowned senior sales and marketing leader with a solid track record of achievements across competitive technology industries. For over twenty years, he has driven business model transformations, go-to-market strategies, and organizational evolution. As an execution excellence evangelist and alumnus of Microsoft, Kaspersky Lab, and Seagate Technology, Frolov has trained teams at all levels on how to deliver tangible results in a complex global business environment, often bucking popular advice and "how we've always done it" processes. *Terrible Advice: Everything You've Been Told about Succeeding during a Recession Is Wrong (And What to Do about It): A Practical Guide to Execution Excellence in Turbulent Times* is his first book. Talk to Maxim Frolov about executing with excellence in your business at www.MaximFrolov.consulting.